4

KUMON MATH WORKBOOKS

Geometry & Measurement

Table of Contents

KUMON

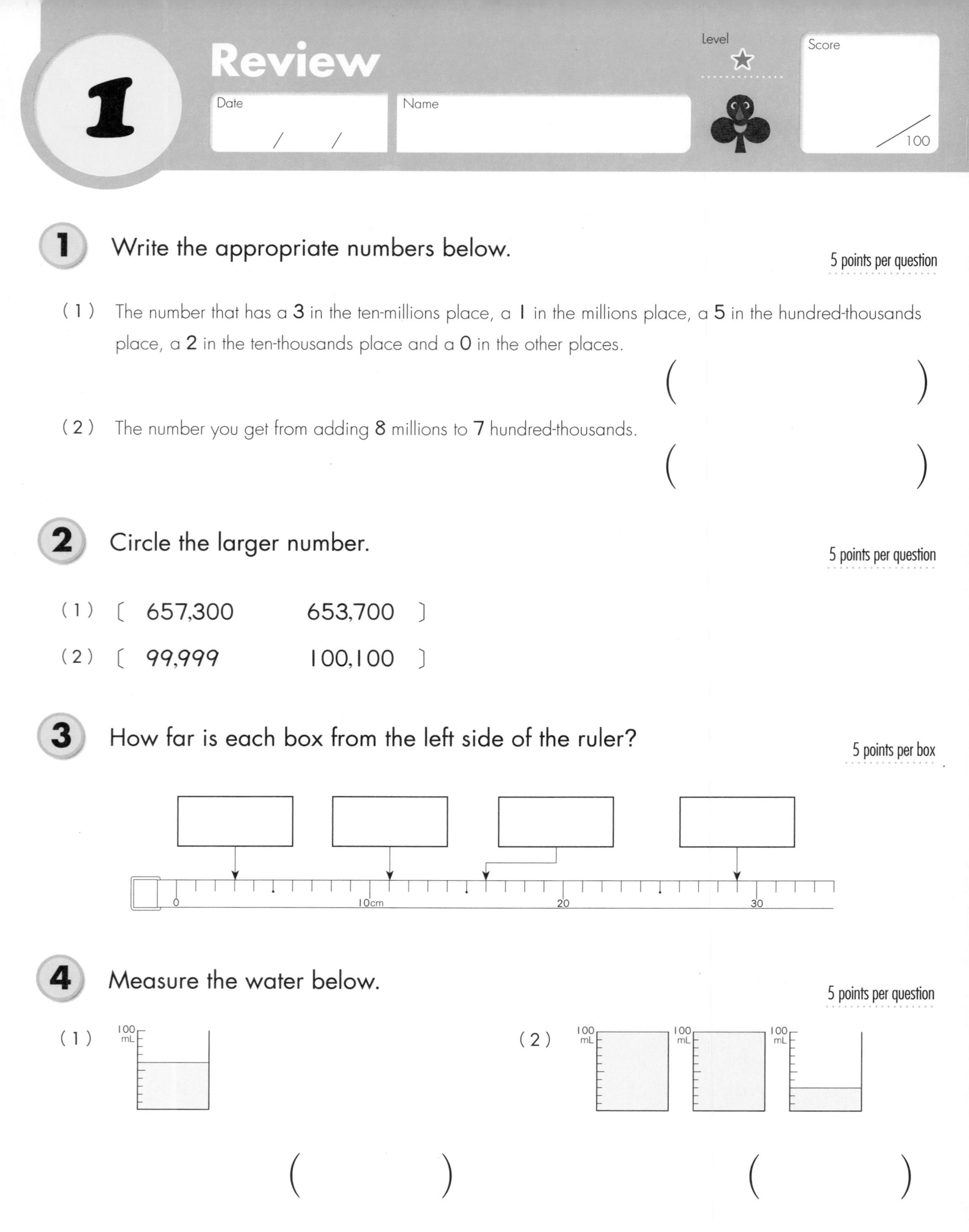

1 Review

Level ☆

Date / /

Name

Score /100

1 Write the appropriate numbers below.

5 points per question

(1) The number that has a **3** in the ten-millions place, a **1** in the millions place, a **5** in the hundred-thousands place, a **2** in the ten-thousands place and a **0** in the other places.

( )

(2) The number you get from adding **8** millions to **7** hundred-thousands.

( )

2 Circle the larger number.

5 points per question

(1) 〔 657,300 653,700 〕

(2) 〔 99,999 100,100 〕

3 How far is each box from the left side of the ruler?

5 points per box

4 Measure the water below.

5 points per question

(1)

( )

(2)

( )

5 Draw the following shapes on the grid below.

10 points per question

a A square with sides that are 3 centimeters long.
b A rectangle with sides that are 2 centimeters and 4 centimeters long.
c A triangle with two lines that connect at a right angle and are 4 centimeters and 5 centimeters long.

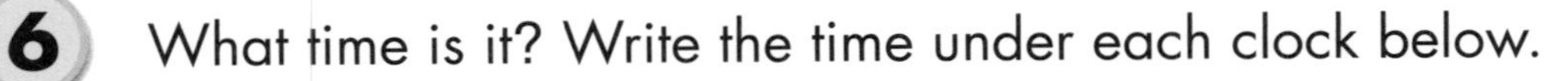

6 What time is it? Write the time under each clock below.

5 points per question

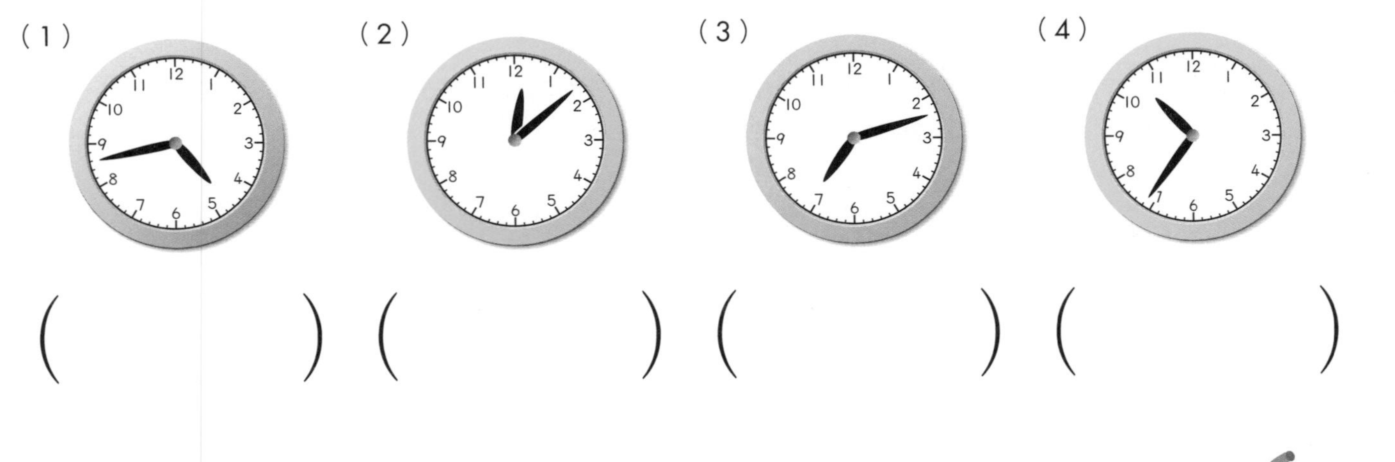

( ) ( ) ( ) ( )

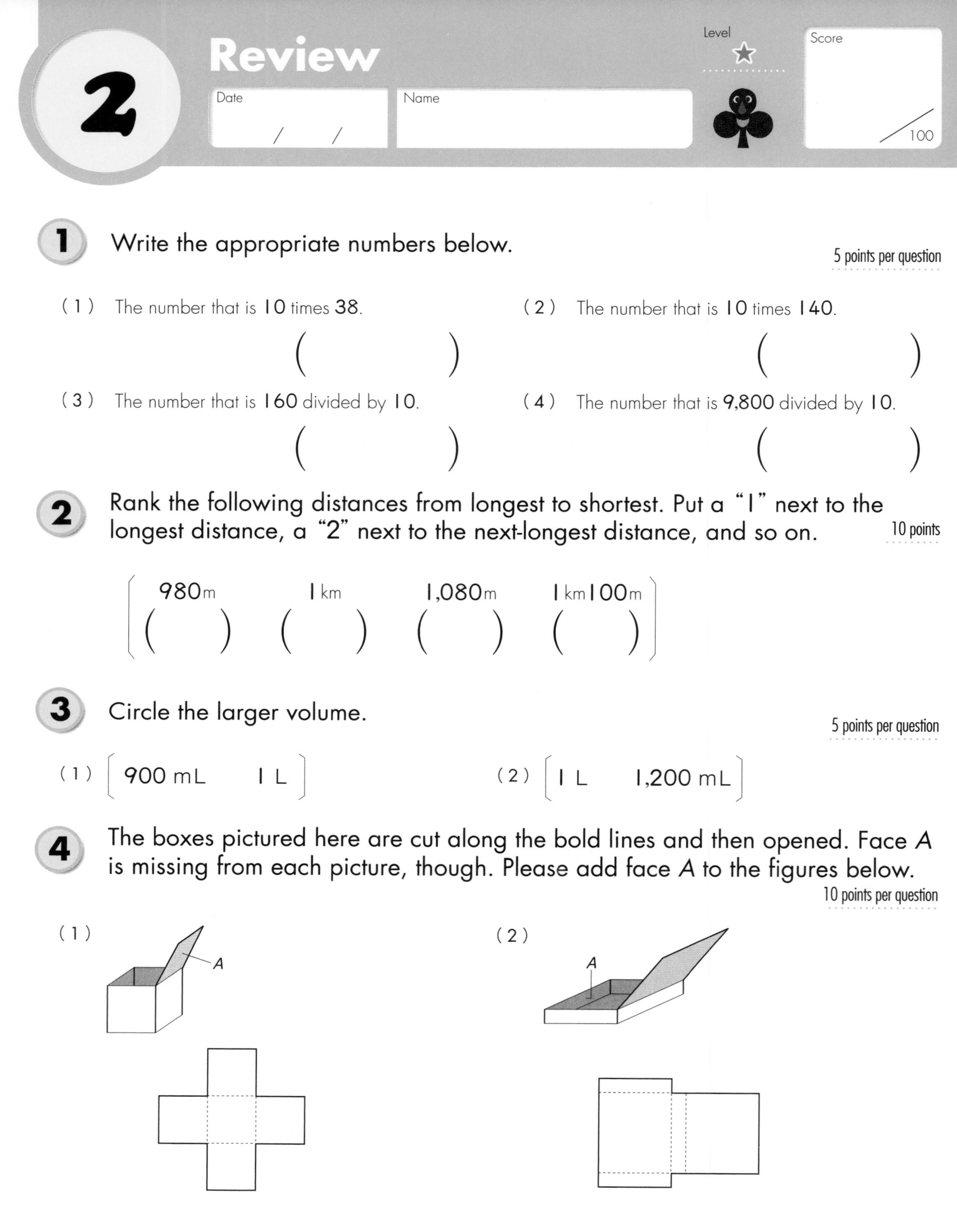

2 Review

Date / / Name Level ☆ Score /100

1 Write the appropriate numbers below. 5 points per question

(1) The number that is 10 times 38. ( )

(2) The number that is 10 times 140. ( )

(3) The number that is 160 divided by 10. ( )

(4) The number that is 9,800 divided by 10. ( )

2 Rank the following distances from longest to shortest. Put a "1" next to the longest distance, a "2" next to the next-longest distance, and so on. 10 points

[980 m ( ) 1 km ( ) 1,080 m ( ) 1 km 100 m ( )]

3 Circle the larger volume. 5 points per question

(1) [900 mL 1 L]

(2) [1 L 1,200 mL]

4 The boxes pictured here are cut along the bold lines and then opened. Face *A* is missing from each picture, though. Please add face *A* to the figures below. 10 points per question

(1) *A*

(2) *A*

5 Read the weight on each scale and then write it below.

5 points per question

(1)

0
2kg
1kg500g
500g
1kg

(   )

(2)

0
2kg
1kg500g
500g
1kg

(   )

6 How many of each type of face do the boxes below have?

5 points per question

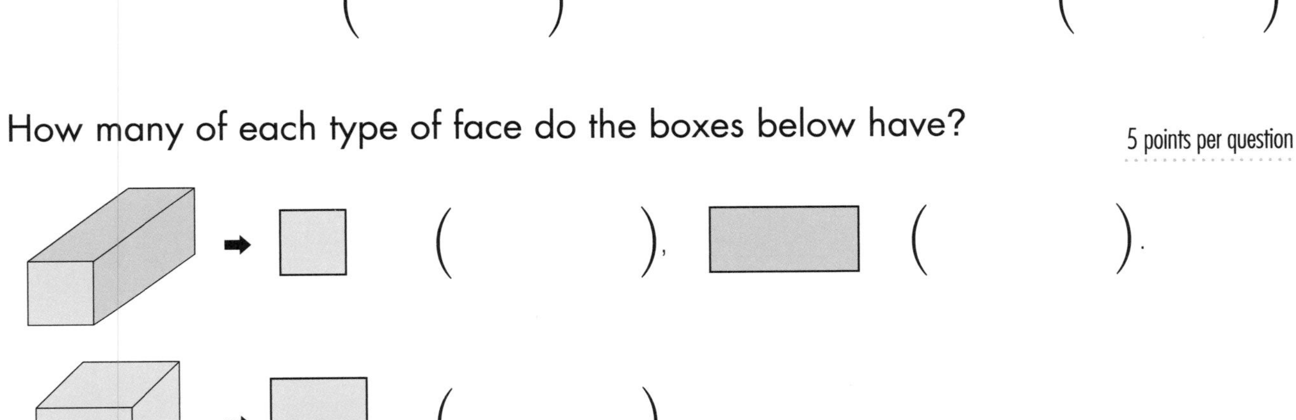

(1) ➡ (   ), (   ).

(2) ➡ (   ).

7 Sort the shapes pictured here into the categories below.

10 points per question

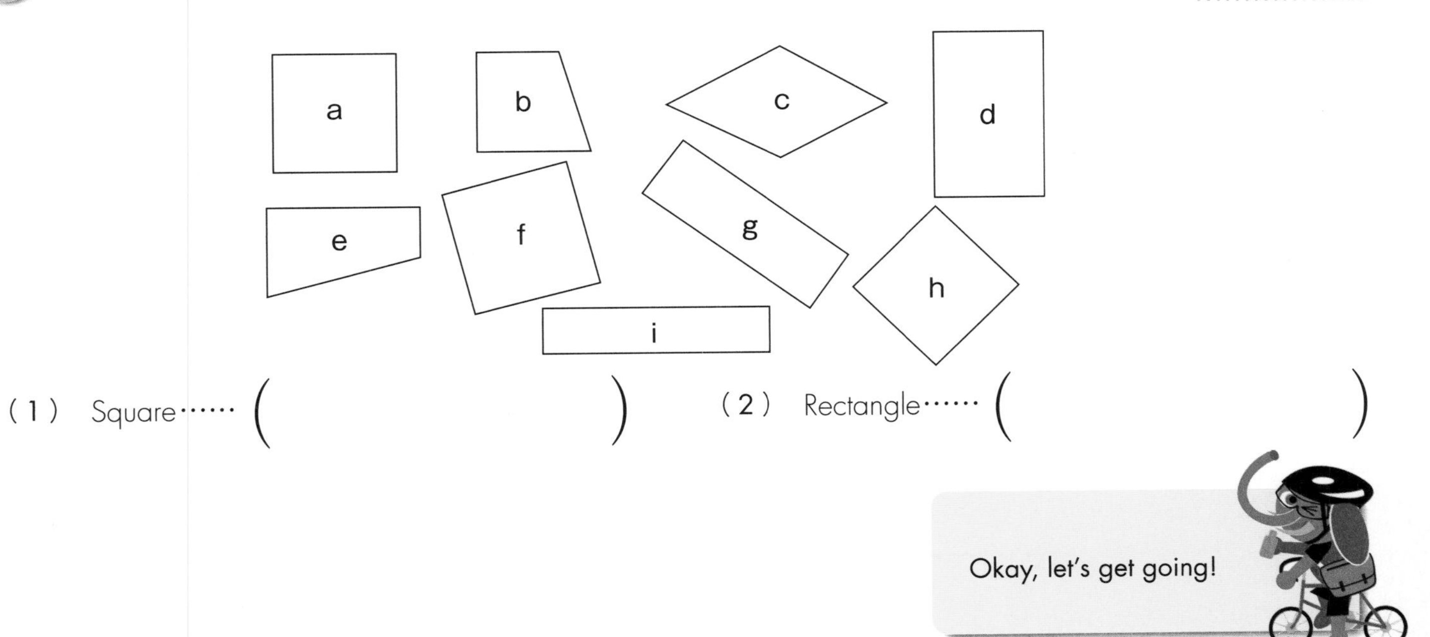

(1) Square······ (   )

(2) Rectangle······ (   )

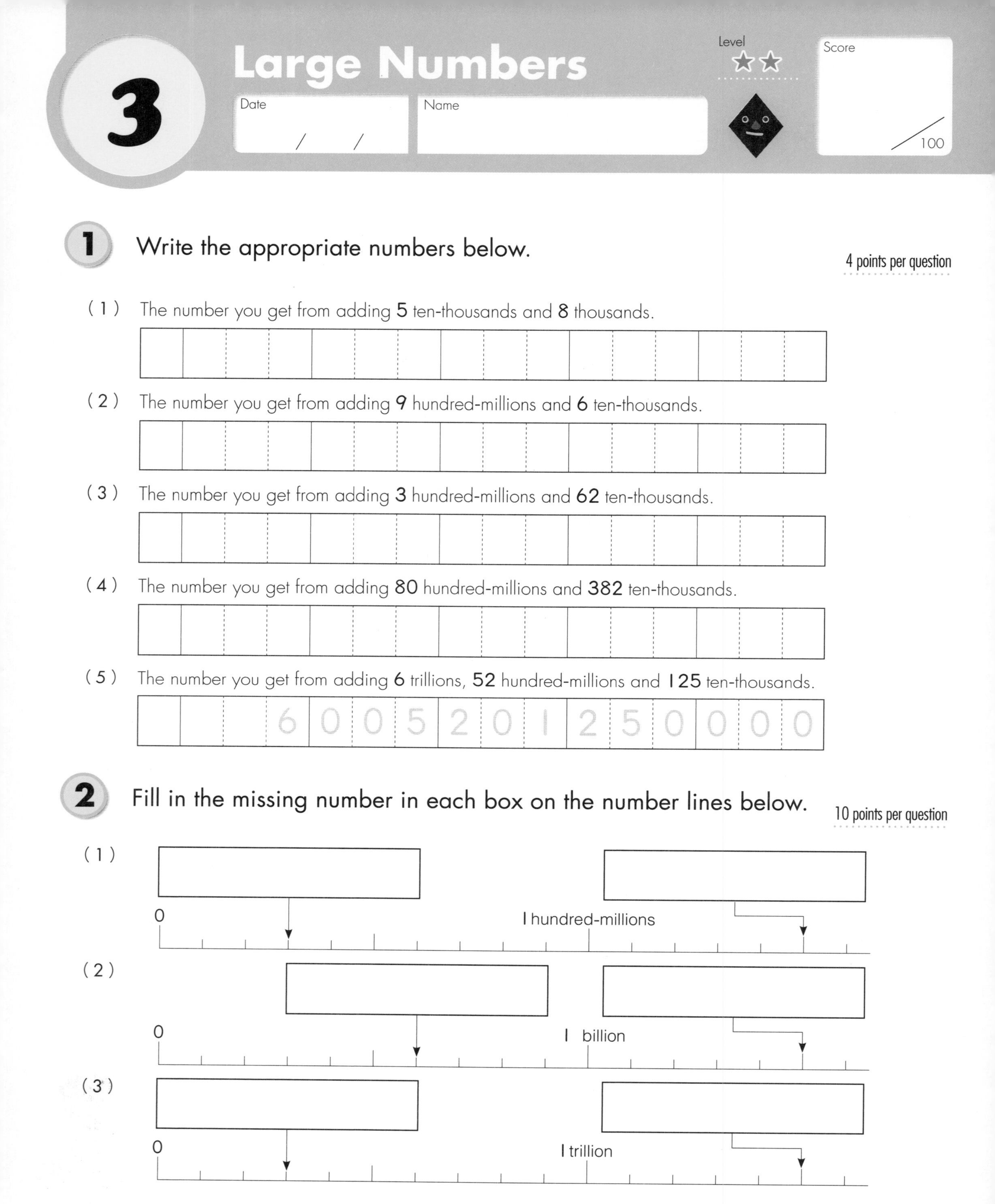

3 Large Numbers

Level ☆☆

Date / /

Name

Score /100

1 Write the appropriate numbers below.

4 points per question

(1) The number you get from adding **5** ten-thousands and **8** thousands.

(2) The number you get from adding **9** hundred-millions and **6** ten-thousands.

(3) The number you get from adding **3** hundred-millions and **62** ten-thousands.

(4) The number you get from adding **80** hundred-millions and **382** ten-thousands.

(5) The number you get from adding **6** trillions, **52** hundred-millions and **125** ten-thousands.

6 0 0 5 2 0 1 2 5 0 0 0 0

2 Fill in the missing number in each box on the number lines below.

10 points per question

(1)

(2)

(3)

3 Write the appropriate numbers below.

10 points per question

(1) What is 10 times and 100 times 100,000,000?

(ones place)

					1	0	0	0	0	0	0	0	0
●10 times				1	0	0	0	0	0	0	0	0	0
●100 times													

10 times, 10 times, 100 times

(2) What is 100,000,000 divided by 10 and 100?

(ones place)

					1	0	0	0	0	0	0	0	0
●Divided by 10						1	0	0	0	0	0	0	0
●Divided by 100													

÷10, ÷10, ÷100

(3) What is 10 times and 100 times 320,000,000?

(ones place)

					3	2	0	0	0	0	0	0	0
●10 times													
●100 times													

(4) What is 320,000,000 divided by 10 and 100?

(ones place)

					3	2	0	0	0	0	0	0	0
●Divided by 10													
●Divided by 100													

(5) What is 4,800,000,000 times 10 and divided by 10?

(ones place)

●10 times													
				4	8	0	0	0	0	0	0	0	0
●Divided by 10													

Those are some big numbers.
Can you handle them? Good!

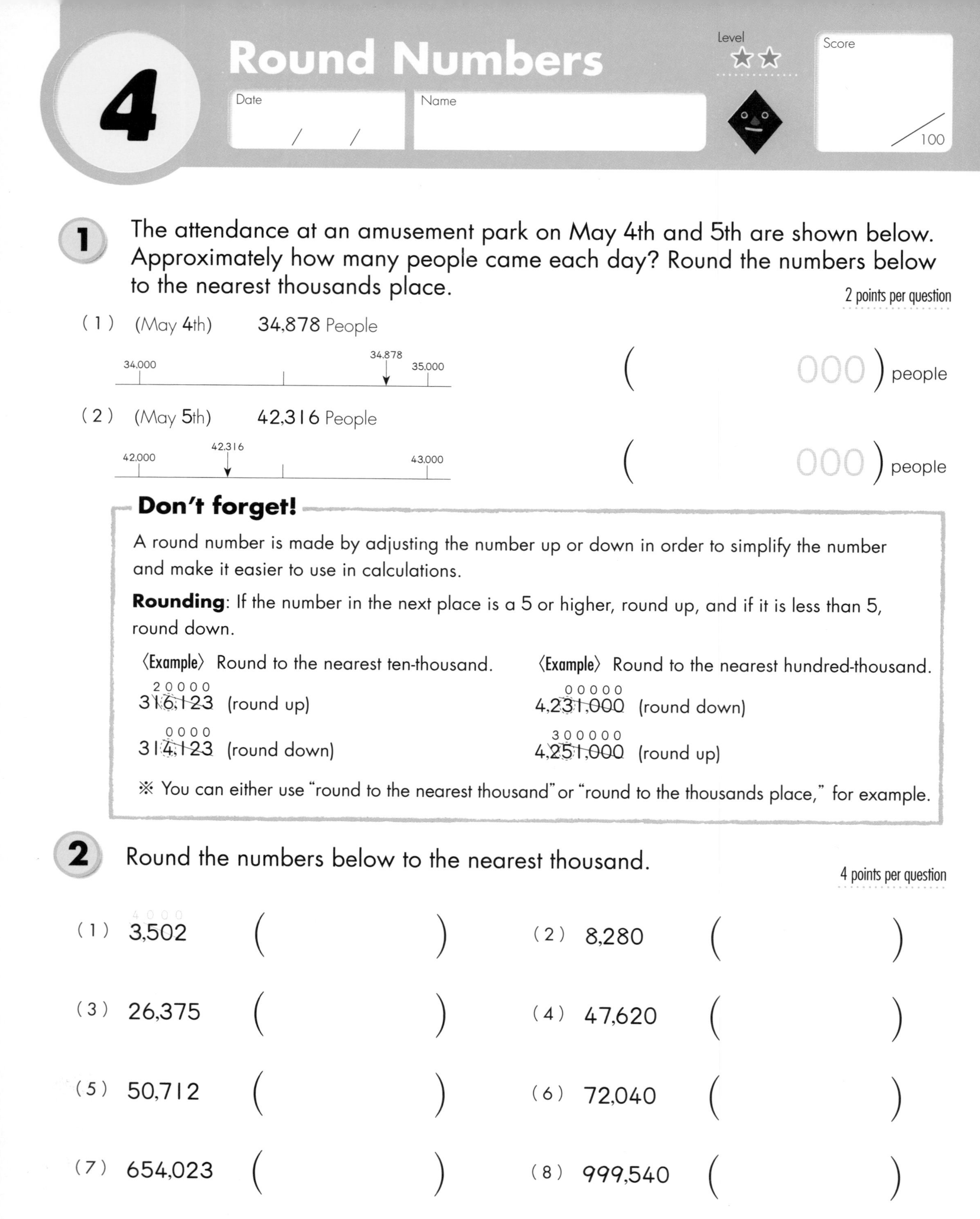

4 Round Numbers

Level ☆☆

Date / /

Name

Score /100

1 The attendance at an amusement park on May 4th and 5th are shown below. Approximately how many people came each day? Round the numbers below to the nearest thousands place.

2 points per question

(1) (May 4th) 34,878 People

(000) people

(2) (May 5th) 42,316 People

(000) people

Don't forget!

A round number is made by adjusting the number up or down in order to simplify the number and make it easier to use in calculations.

Rounding: If the number in the next place is a 5 or higher, round up, and if it is less than 5, round down.

〈Example〉 Round to the nearest ten-thousand.

20000
316,123 (round up)

0000
314,123 (round down)

〈Example〉 Round to the nearest hundred-thousand.

00000
4,231,000 (round down)

300000
4,251,000 (round up)

※ You can either use "round to the nearest thousand" or "round to the thousands place," for example.

2 Round the numbers below to the nearest thousand.

4 points per question

(1) 3,502 ( )

(2) 8,280 ( )

(3) 26,375 ( )

(4) 47,620 ( )

(5) 50,712 ( )

(6) 72,040 ( )

(7) 654,023 ( )

(8) 999,540 ( )

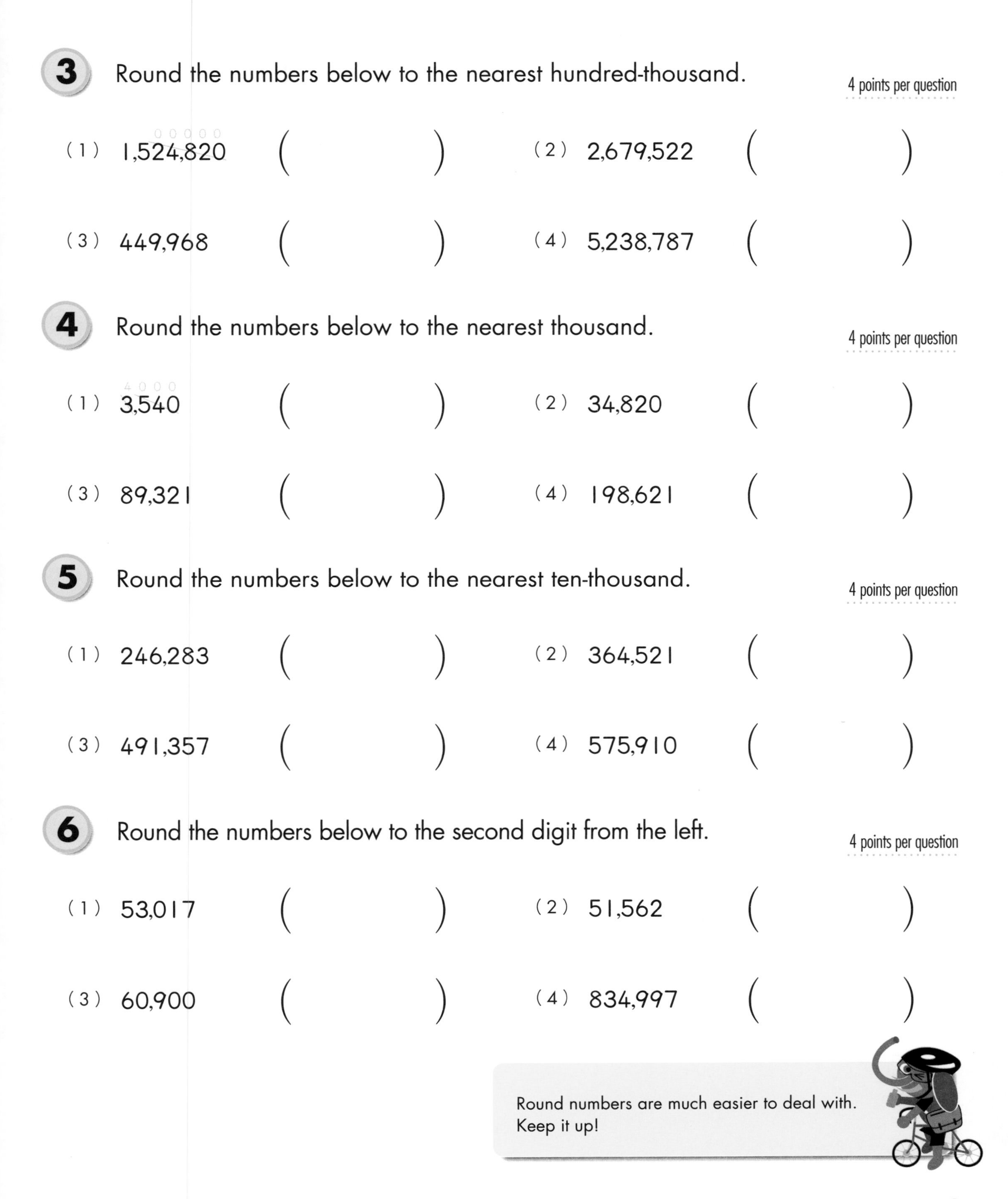

3 Round the numbers below to the nearest hundred-thousand.

4 points per question

(1) 1,524,820 (　　　　)　　(2) 2,679,522 (　　　　)

(3) 449,968 (　　　　)　　(4) 5,238,787 (　　　　)

4 Round the numbers below to the nearest thousand.

4 points per question

(1) 3,540 (　　　　)　　(2) 34,820 (　　　　)

(3) 89,321 (　　　　)　　(4) 198,621 (　　　　)

5 Round the numbers below to the nearest ten-thousand.

4 points per question

(1) 246,283 (　　　　)　　(2) 364,521 (　　　　)

(3) 491,357 (　　　　)　　(4) 575,910 (　　　　)

6 Round the numbers below to the second digit from the left.

4 points per question

(1) 53,017 (　　　　)　　(2) 51,562 (　　　　)

(3) 60,900 (　　　　)　　(4) 834,997 (　　　　)

Round numbers are much easier to deal with.
Keep it up!

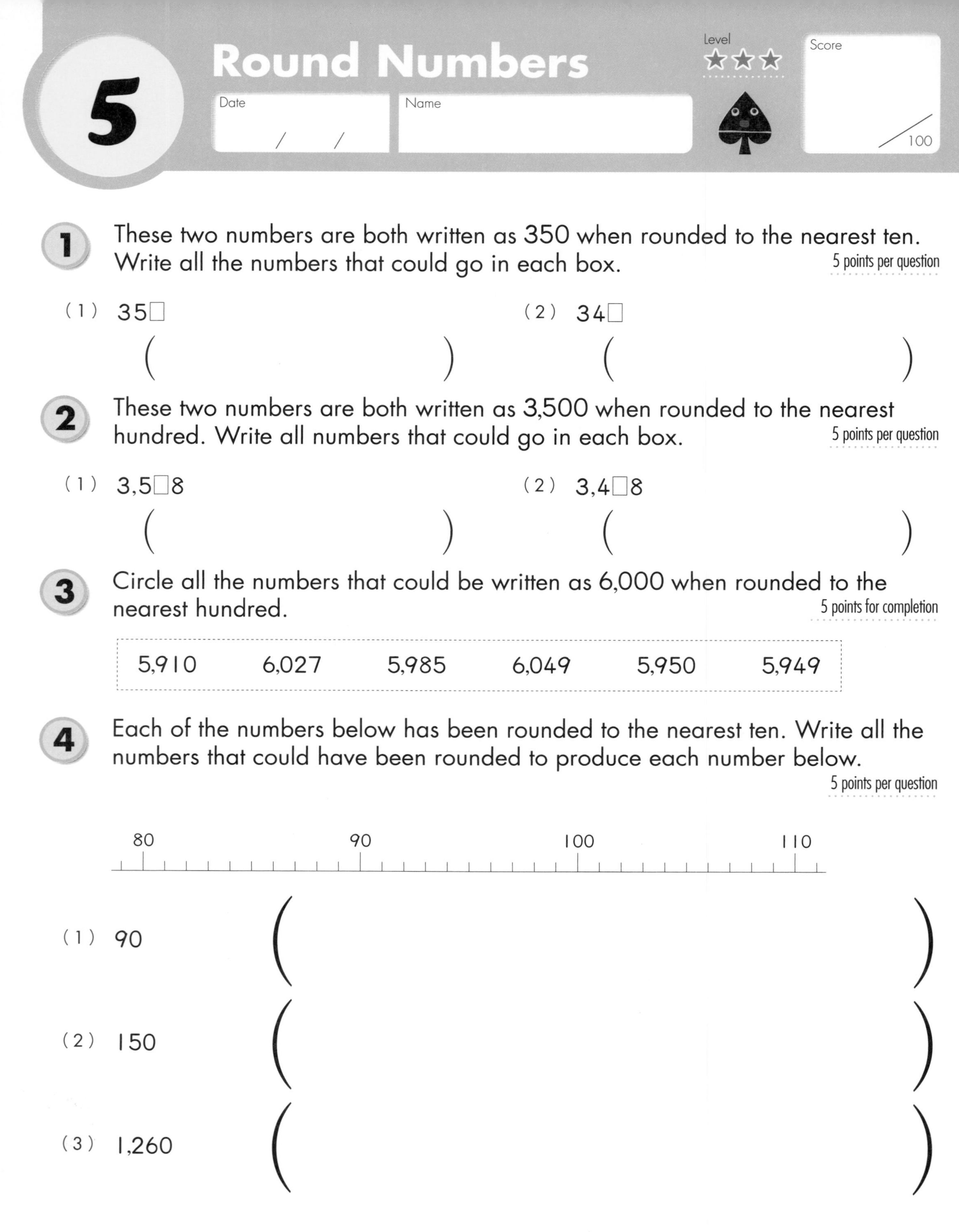

5 Round Numbers

Level ☆☆☆

Date / /

Name

Score /100

1 These two numbers are both written as 350 when rounded to the nearest ten. Write all the numbers that could go in each box.

5 points per question

(1) 35□

( )

(2) 34□

( )

2 These two numbers are both written as 3,500 when rounded to the nearest hundred. Write all numbers that could go in each box.

5 points per question

(1) 3,5□8

( )

(2) 3,4□8

( )

3 Circle all the numbers that could be written as 6,000 when rounded to the nearest hundred.

5 points for completion

5,910	6,027	5,985	6,049	5,950	5,949

4 Each of the numbers below has been rounded to the nearest ten. Write all the numbers that could have been rounded to produce each number below.

5 points per question

80 90 100 110

(1) 90 ( )

(2) 150 ( )

(3) 1,260 ( )

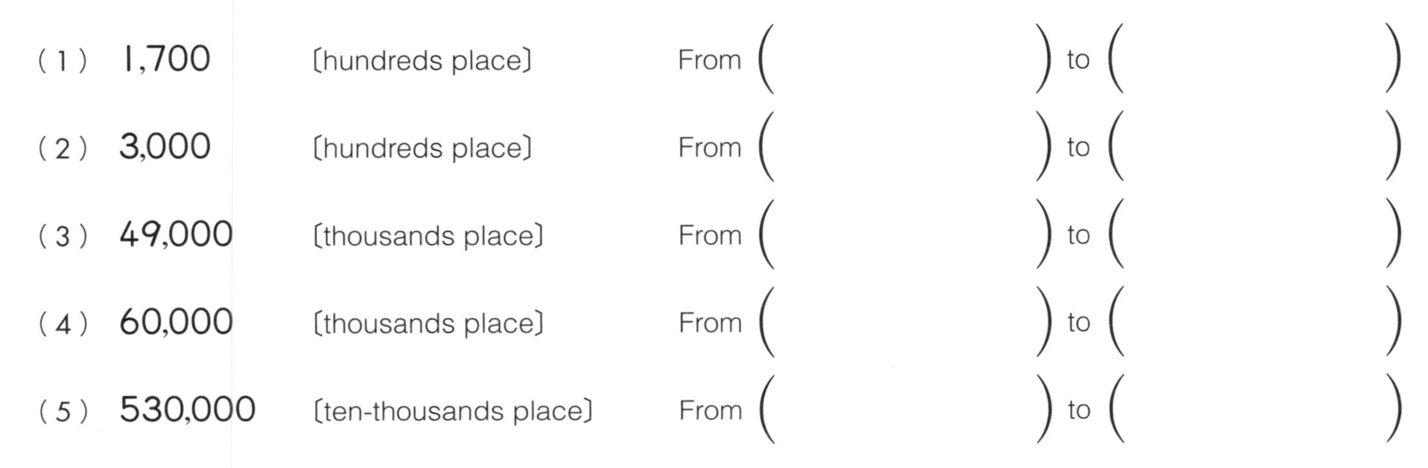

5 Each of the numbers below has been rounded. Write all the numbers that could have been rounded as shown to produce the numbers in each case below.

5 points per question

(1) 1,700 〔hundreds place〕 From ( ) to ( )

(2) 3,000 〔hundreds place〕 From ( ) to ( )

(3) 49,000 〔thousands place〕 From ( ) to ( )

(4) 60,000 〔thousands place〕 From ( ) to ( )

(5) 530,000 〔ten-thousands place〕 From ( ) to ( )

6 Each of the numbers below has been rounded to the second digit from the left. Write all the possible numbers that could have been rounded to produce each number below.

5 points per question

(1) 890 From ( ) to ( )

(2) 4,300 From ( ) to ( )

(3) 9,100 From ( ) to ( )

(4) 15,000 From ( ) to ( )

(5) 72,000 From ( ) to ( )

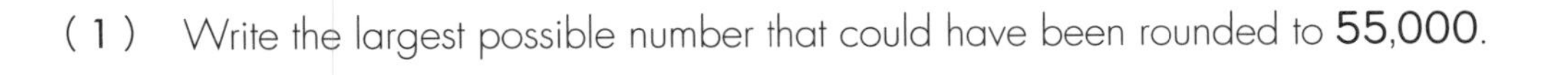

7 The following questions are about numbers that have been rounded to their nearest thousand in order to produce 55,000.

5 points per question

(1) Write the largest possible number that could have been rounded to 55,000. ( )

(2) Write the smallest possible number that could have been rounded to 55,000. ( )

You're doing great. Good job!

1 Round the numbers below to the nearest hundred in order to complete each calculation.

4 points per question

(1) 1,240 + 3,568

1,200 + 3,600 =

(2) 2,087 + 326

2,100 + 300 =

(3) 3,472 + 8,735

(4) 452 + 8,117

(5) 8,608 − 6,594

(6) 7,721 − 954

(7) 6,890 − 4,683

(8) 5,667 − 875

2 Round the numbers below to the nearest thousand in order to complete each calculation.

4 points per question

(1) 34,326 + 25,587

34,000 + 26,000 =

(2) 18,752 + 3,624

(3) 75,819 + 36,490

(4) 5,921 + 86,418

(5) 98,107 − 76,805

(6) 88,505 − 9,104

(7) 7,608 + 3,452 + 4,548

(8) 8,953 − 2,421 − 3,784

3 Round the numbers below to the nearest ten-thousand in order to complete each calculation.

3 points per question

(1) 62,543 + 36,875

(2) 49,803 + 222,675

(3) 96,873 − 50,032

(4) 572,156 − 431,890

(5) 680,289 + 4,569,741 − 58,040

4 Round the numbers below as indicated in order to complete each calculation.

3 points per question

(1) 〔tens place〕

348 + 82

(2) 〔hundreds place〕

352 + 3,135

(3) 〔hundreds place〕

4,513 + 6,462

(4) 〔thousands place〕

34,910 + 8,294

(5) 〔thousands place〕

73,404 − 52,783

(6) 〔ten-thousands place〕

149,321 + 323,584

(7) 〔ten-thousands place〕

386,304 − 257,392 + 523,608

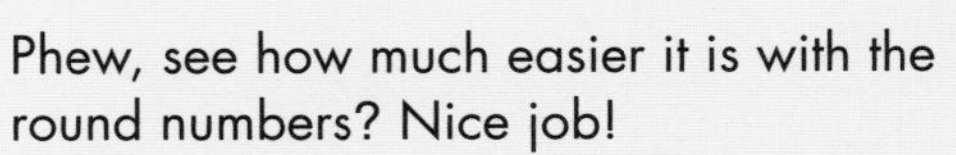

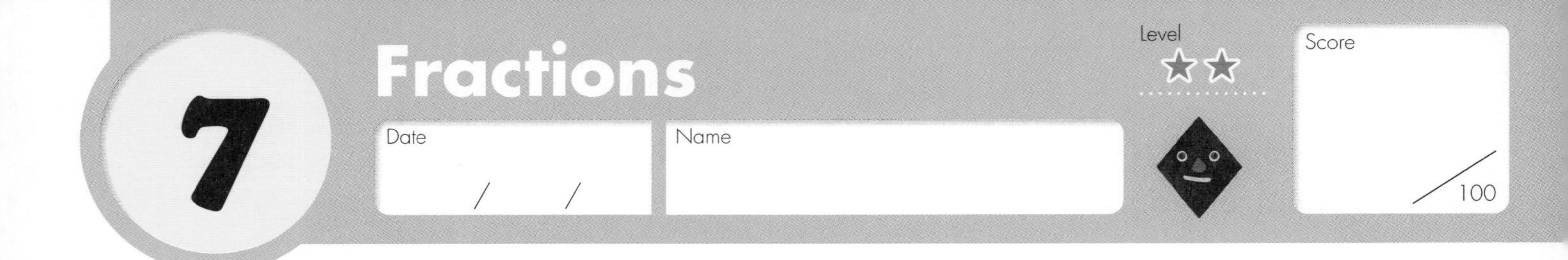

7 Fractions

Level ☆☆

Date / /

Name

Score /100

1 The meter below is cut into 5 equal parts. Write the appropriate numbers in each box below.

5 points per question

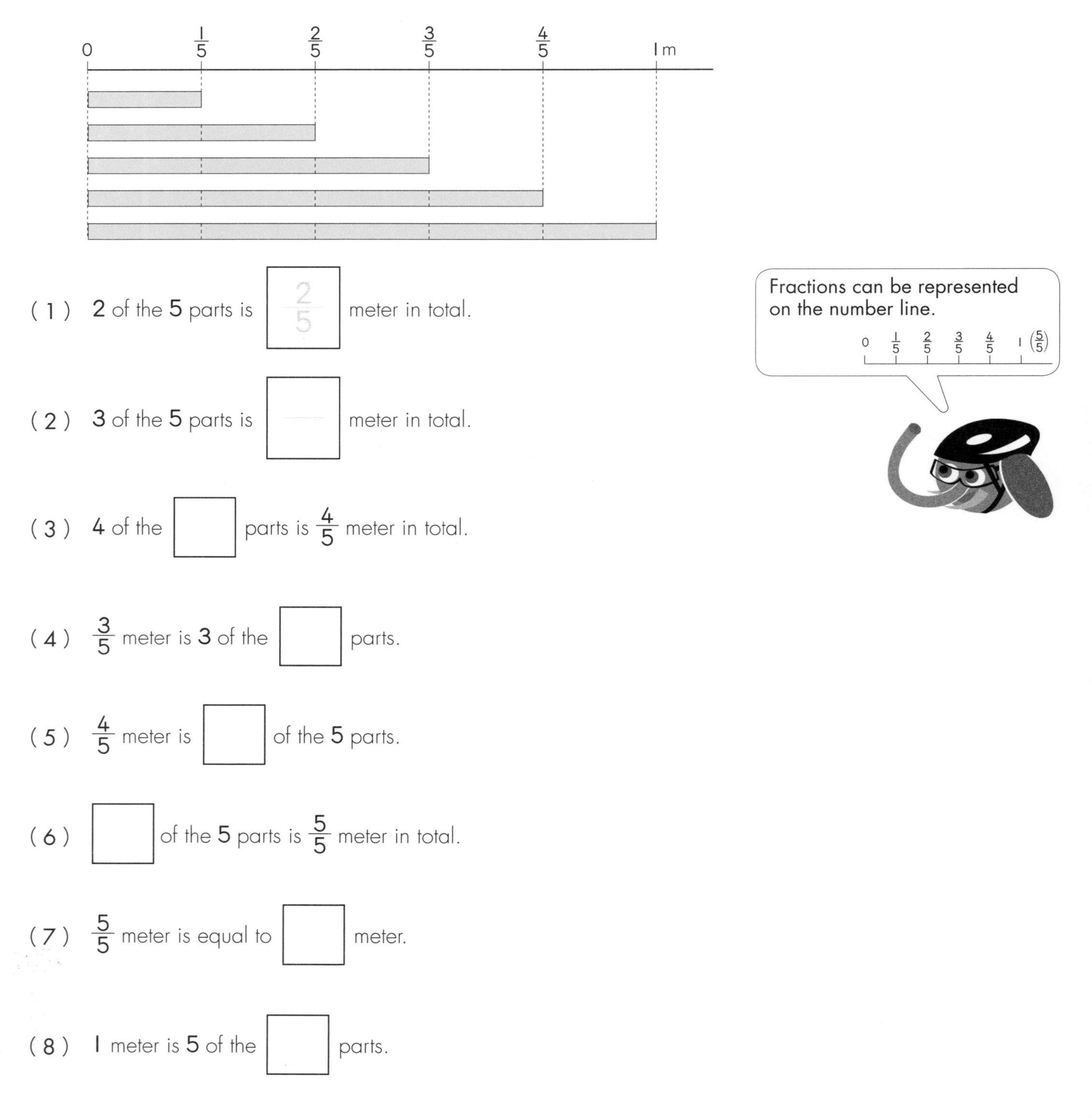

(1) 2 of the 5 parts is $\frac{2}{5}$ meter in total.

(2) 3 of the 5 parts is ☐ meter in total.

(3) 4 of the ☐ parts is $\frac{4}{5}$ meter in total.

(4) $\frac{3}{5}$ meter is 3 of the ☐ parts.

(5) $\frac{4}{5}$ meter is ☐ of the 5 parts.

(6) ☐ of the 5 parts is $\frac{5}{5}$ meter in total.

(7) $\frac{5}{5}$ meter is equal to ☐ meter.

(8) 1 meter is 5 of the ☐ parts.

2 Write the appropriate fraction in each box on the number lines below.

5 points per box

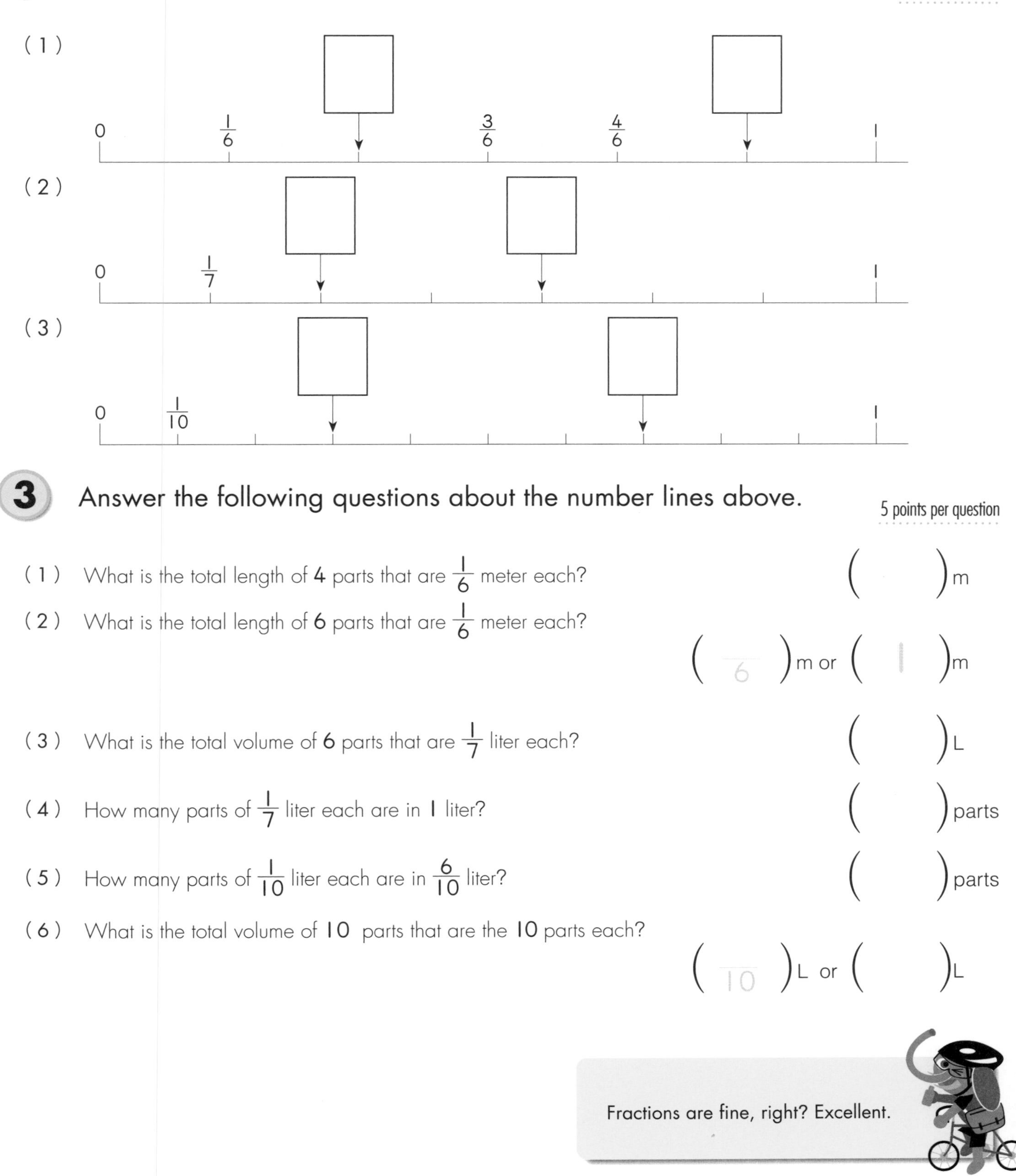

3 Answer the following questions about the number lines above.

5 points per question

(1) What is the total length of **4** parts that are $\frac{1}{6}$ meter each? ( )m

(2) What is the total length of **6** parts that are $\frac{1}{6}$ meter each? ($\frac{}{6}$)m or (1)m

(3) What is the total volume of **6** parts that are $\frac{1}{7}$ liter each? ( )L

(4) How many parts of $\frac{1}{7}$ liter each are in 1 liter? ( )parts

(5) How many parts of $\frac{1}{10}$ liter each are in $\frac{6}{10}$ liter? ( )parts

(6) What is the total volume of 10 parts that are the 10 parts each? ($\frac{}{10}$)L or ( )L

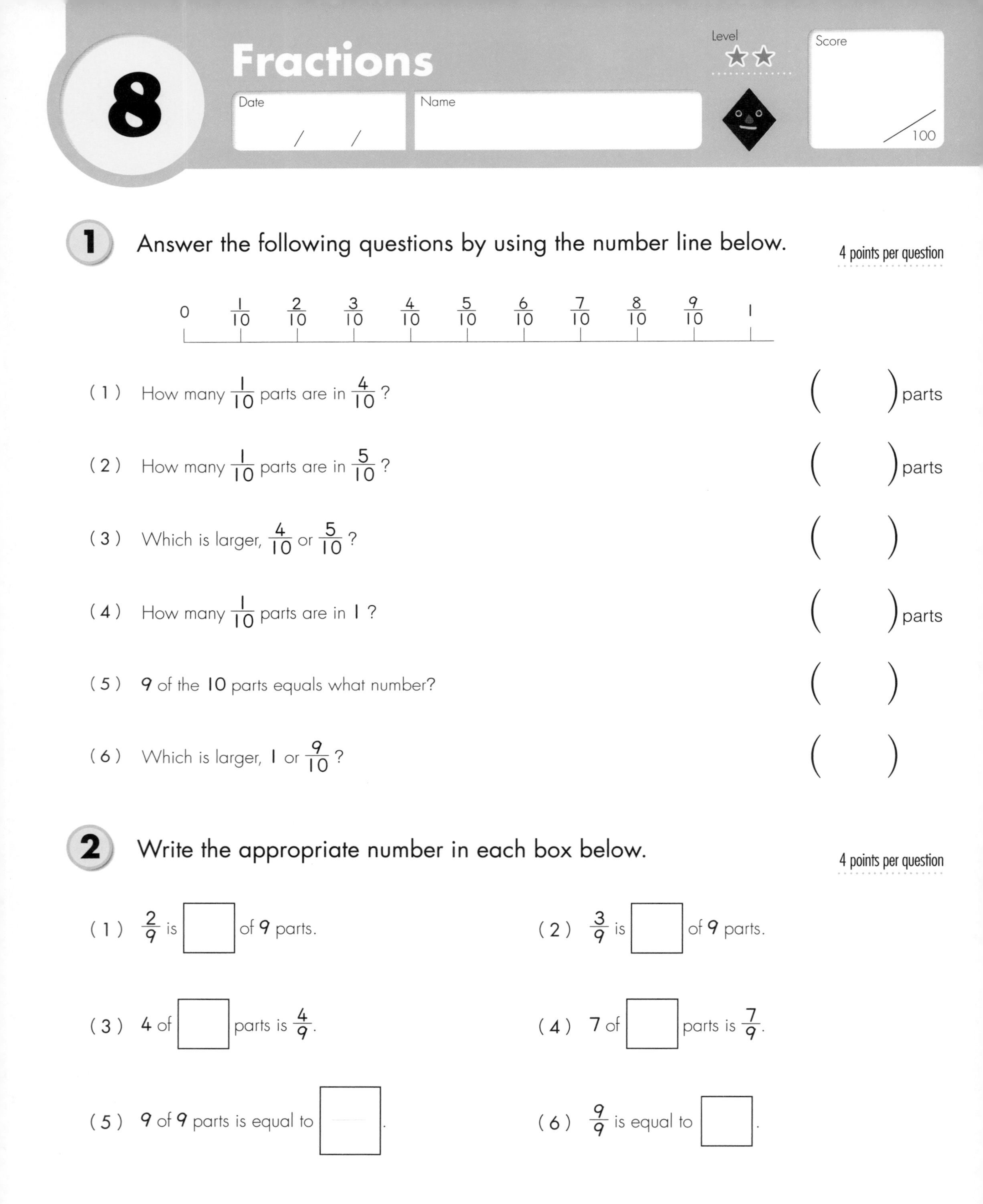

8 Fractions

Level ☆☆

Date / /

Name

Score /100

1 Answer the following questions by using the number line below.

4 points per question

$0 \quad \frac{1}{10} \quad \frac{2}{10} \quad \frac{3}{10} \quad \frac{4}{10} \quad \frac{5}{10} \quad \frac{6}{10} \quad \frac{7}{10} \quad \frac{8}{10} \quad \frac{9}{10} \quad 1$

(1) How many $\frac{1}{10}$ parts are in $\frac{4}{10}$? ( ) parts

(2) How many $\frac{1}{10}$ parts are in $\frac{5}{10}$? ( ) parts

(3) Which is larger, $\frac{4}{10}$ or $\frac{5}{10}$? ( )

(4) How many $\frac{1}{10}$ parts are in 1 ? ( ) parts

(5) 9 of the 10 parts equals what number? ( )

(6) Which is larger, 1 or $\frac{9}{10}$? ( )

2 Write the appropriate number in each box below.

4 points per question

(1) $\frac{2}{9}$ is ☐ of 9 parts.

(2) $\frac{3}{9}$ is ☐ of 9 parts.

(3) 4 of ☐ parts is $\frac{4}{9}$.

(4) 7 of ☐ parts is $\frac{7}{9}$.

(5) 9 of 9 parts is equal to ☐.

(6) $\frac{9}{9}$ is equal to ☐.

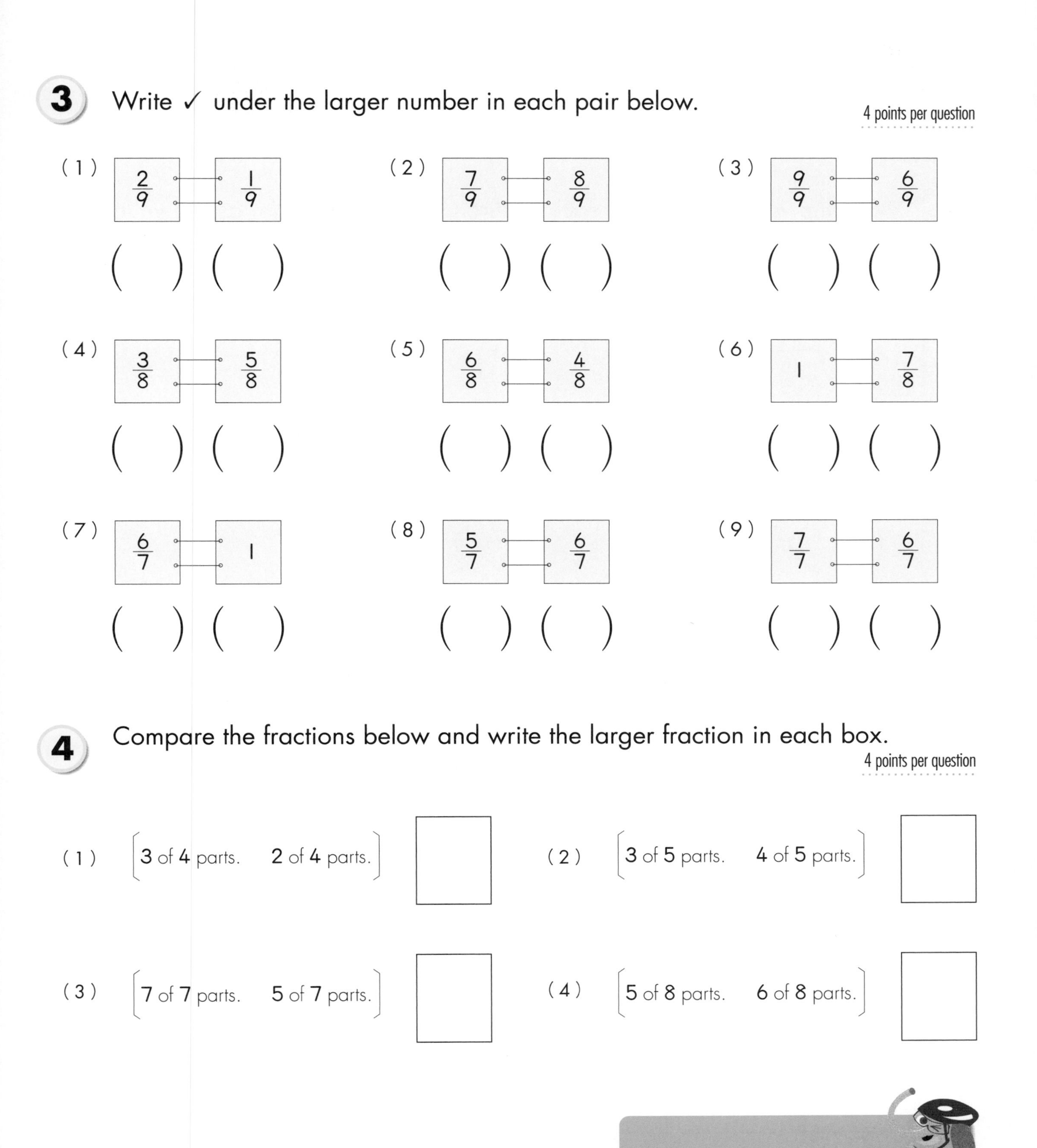

3 Write ✓ under the larger number in each pair below.

4 points per question

(1) $\frac{2}{9}$ $\frac{1}{9}$ ( ) ( )

(2) $\frac{7}{9}$ $\frac{8}{9}$ ( ) ( )

(3) $\frac{9}{9}$ $\frac{6}{9}$ ( ) ( )

(4) $\frac{3}{8}$ $\frac{5}{8}$ ( ) ( )

(5) $\frac{6}{8}$ $\frac{4}{8}$ ( ) ( )

(6) 1 $\frac{7}{8}$ ( ) ( )

(7) $\frac{6}{7}$ 1 ( ) ( )

(8) $\frac{5}{7}$ $\frac{6}{7}$ ( ) ( )

(9) $\frac{7}{7}$ $\frac{6}{7}$ ( ) ( )

4 Compare the fractions below and write the larger fraction in each box.

4 points per question

(1) [3 of 4 parts. 2 of 4 parts.] □

(2) [3 of 5 parts. 4 of 5 parts.] □

(3) [7 of 7 parts. 5 of 7 parts.] □

(4) [5 of 8 parts. 6 of 8 parts.] □

How's it going? Good I hope!

9 Fractions

Level ★★

Date / / Name

Score /100

1 Write the volume shown in each picture two different ways, as shown in the sample below.

6 points per question

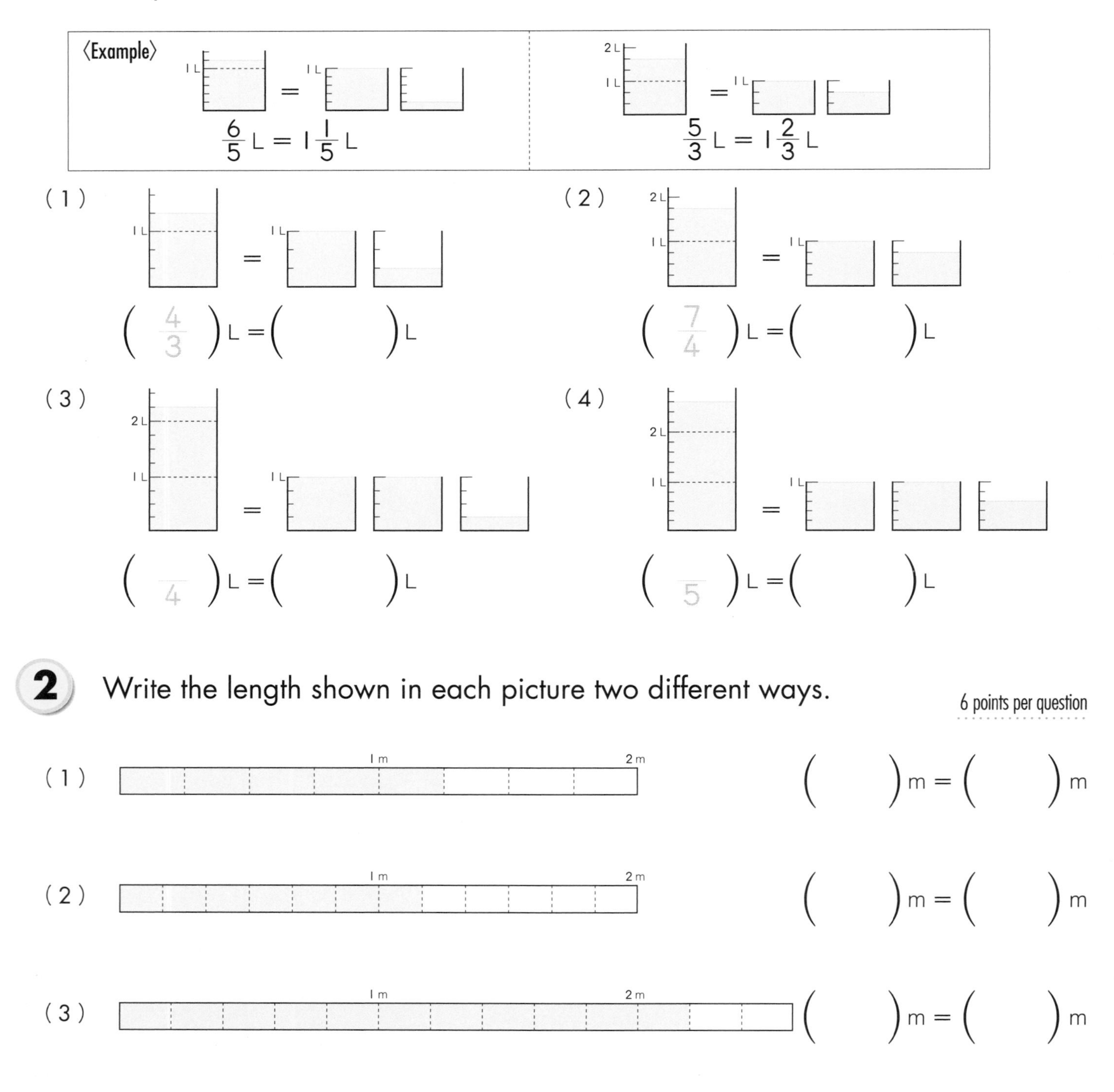

2 Write the length shown in each picture two different ways.

6 points per question

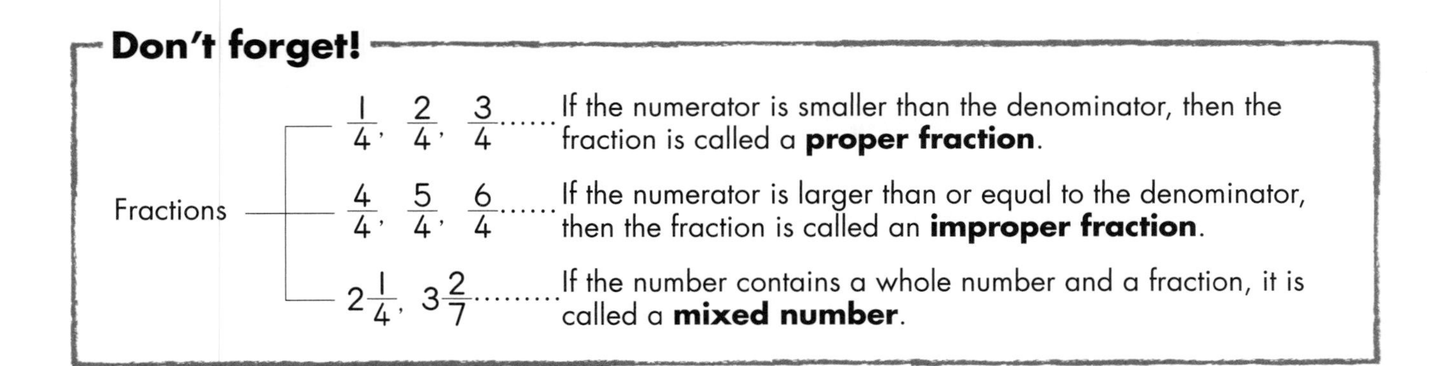

Don't forget!

Fractions

- $\frac{1}{4}, \frac{2}{4}, \frac{3}{4}$ …… If the numerator is smaller than the denominator, then the fraction is called a **proper fraction**.
- $\frac{4}{4}, \frac{5}{4}, \frac{6}{4}$ …… If the numerator is larger than or equal to the denominator, then the fraction is called an **improper fraction**.
- $2\frac{1}{4}, 3\frac{2}{7}$ …… If the number contains a whole number and a fraction, it is called a **mixed number**.

3 Sort the fractions below into proper fractions, improper fractions, and mixed numbers.

10 points for completion

$\frac{7}{6}, \frac{6}{11}, 1\frac{2}{3}, \frac{13}{18}, \frac{15}{15}, \frac{20}{23}, 3\frac{15}{17}, \frac{23}{18}, 4\frac{13}{25}$

Proper fractions (　　　　　　　　)　Improper fractions (　　　　　　　　)

Mixed numbers (　　　　　　　　)

4 Write the appropriate proper and improper fraction in each box on the number lines below.

6 points per box

(1)

(2)

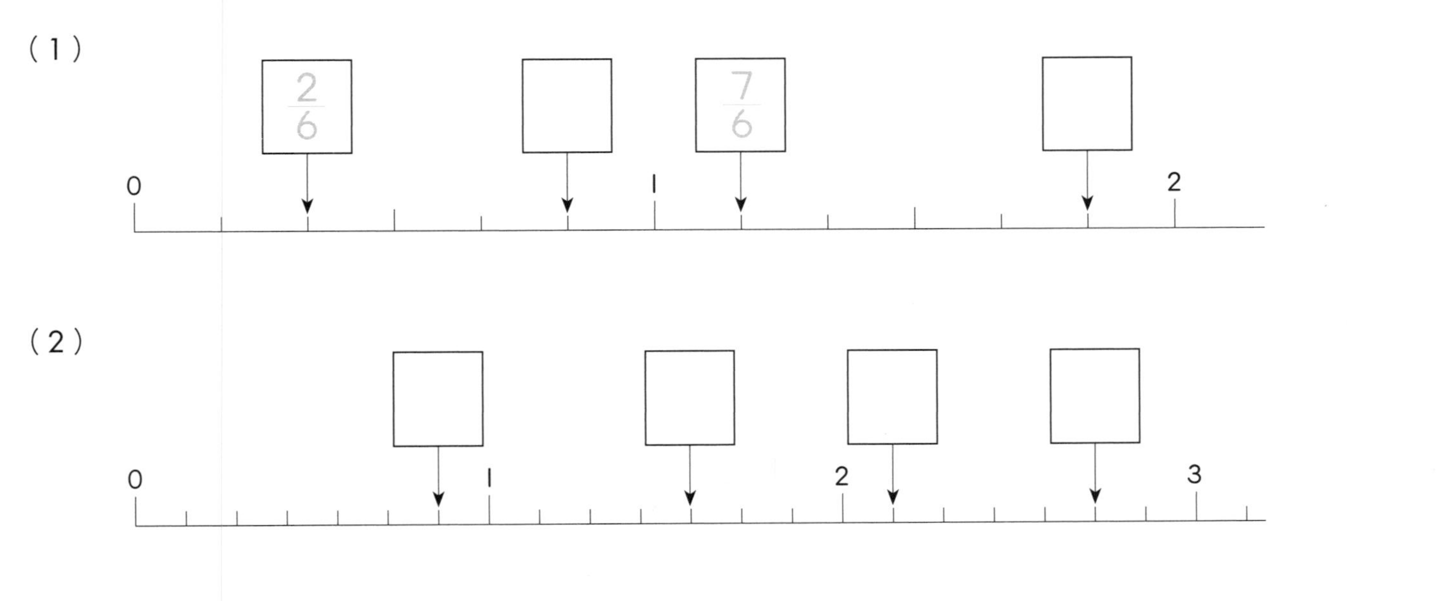

I'm an improper fraction! Just kidding, keep up the good work!

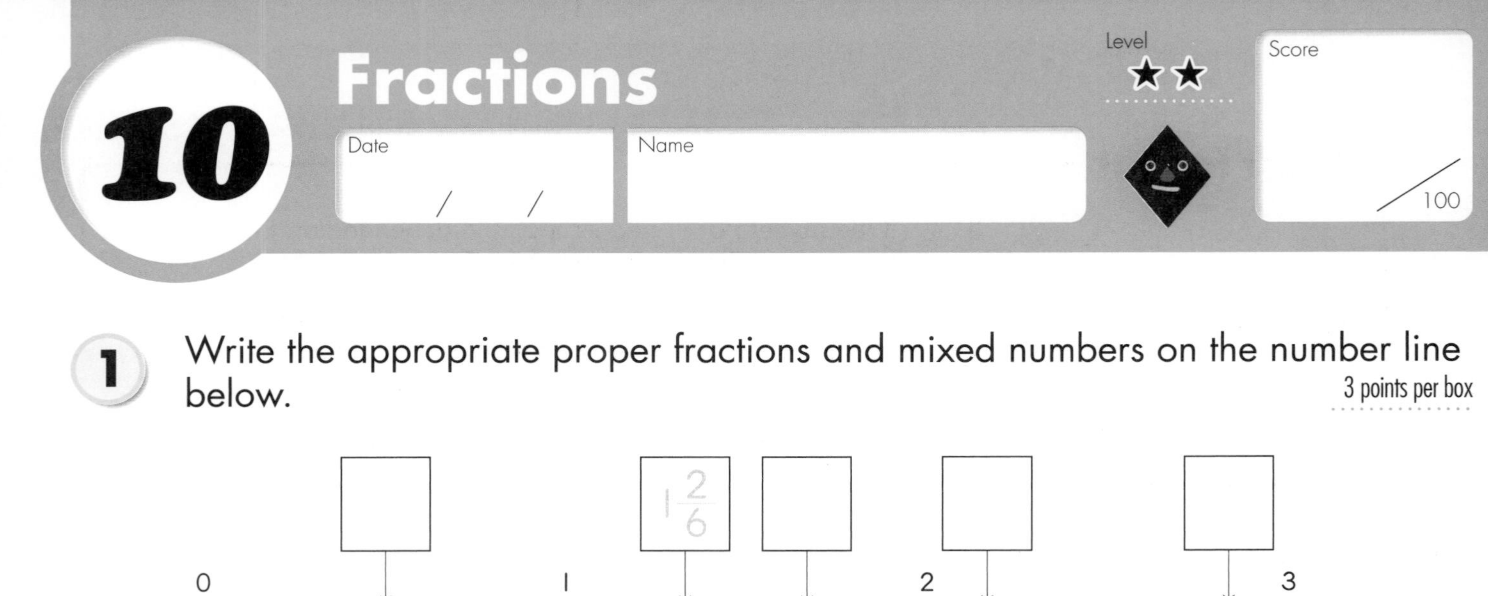

1 Write the appropriate proper fractions and mixed numbers on the number line below.

3 points per box

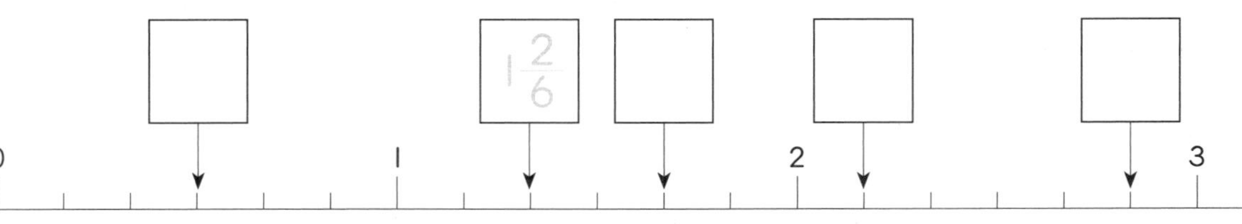

2 Write the appropriate number in each box below.

3 points per box

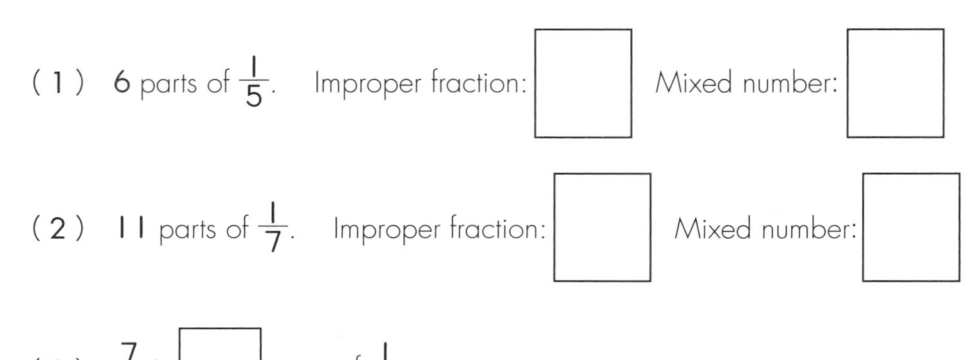

(1) 6 parts of $\frac{1}{5}$. Improper fraction: ☐ Mixed number: ☐

(2) 11 parts of $\frac{1}{7}$. Improper fraction: ☐ Mixed number: ☐

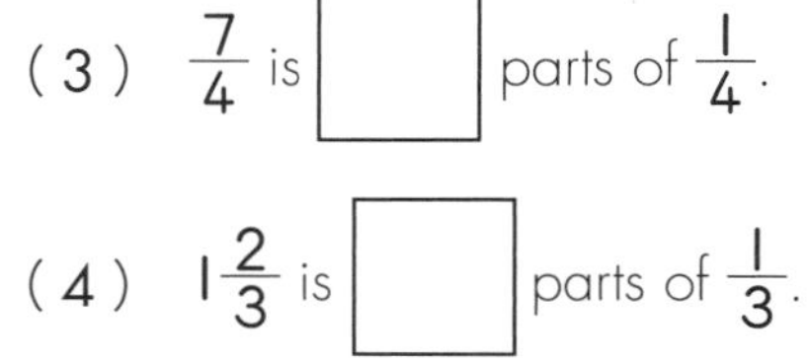

(3) $\frac{7}{4}$ is ☐ parts of $\frac{1}{4}$.

(4) $1\frac{2}{3}$ is ☐ parts of $\frac{1}{3}$.

3 Write ✓ under the larger fraction.

3 points per question

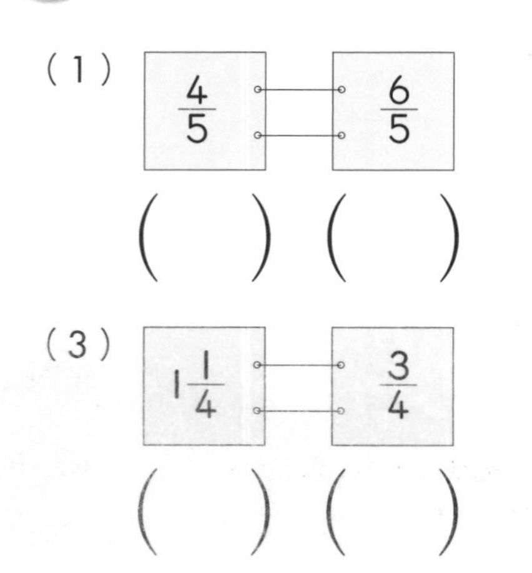

(1) $\frac{4}{5}$ ( ) $\frac{6}{5}$ ( )

(3) $1\frac{1}{4}$ ( ) $\frac{3}{4}$ ( )

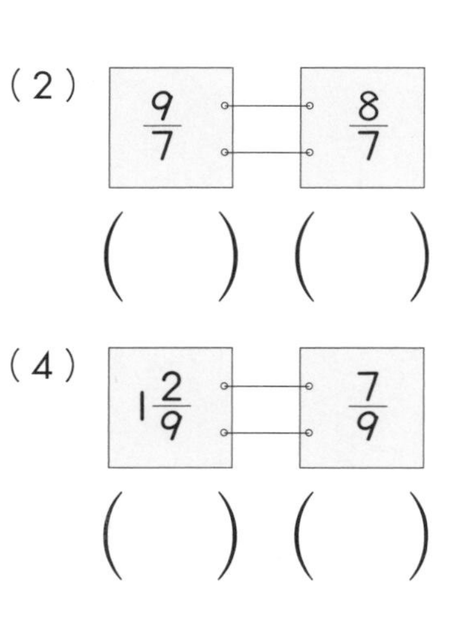

(2) $\frac{9}{7}$ ( ) $\frac{8}{7}$ ( )

(4) $1\frac{2}{9}$ ( ) $\frac{7}{9}$ ( )

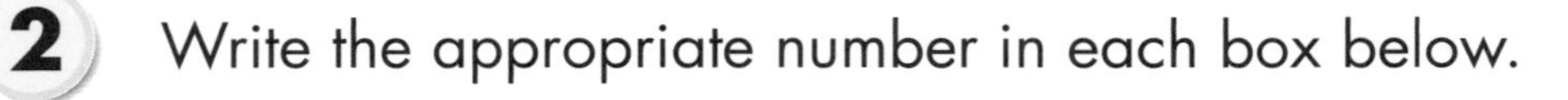

4 Convert each improper fraction below into a mixed number or a whole number.

3 points per question

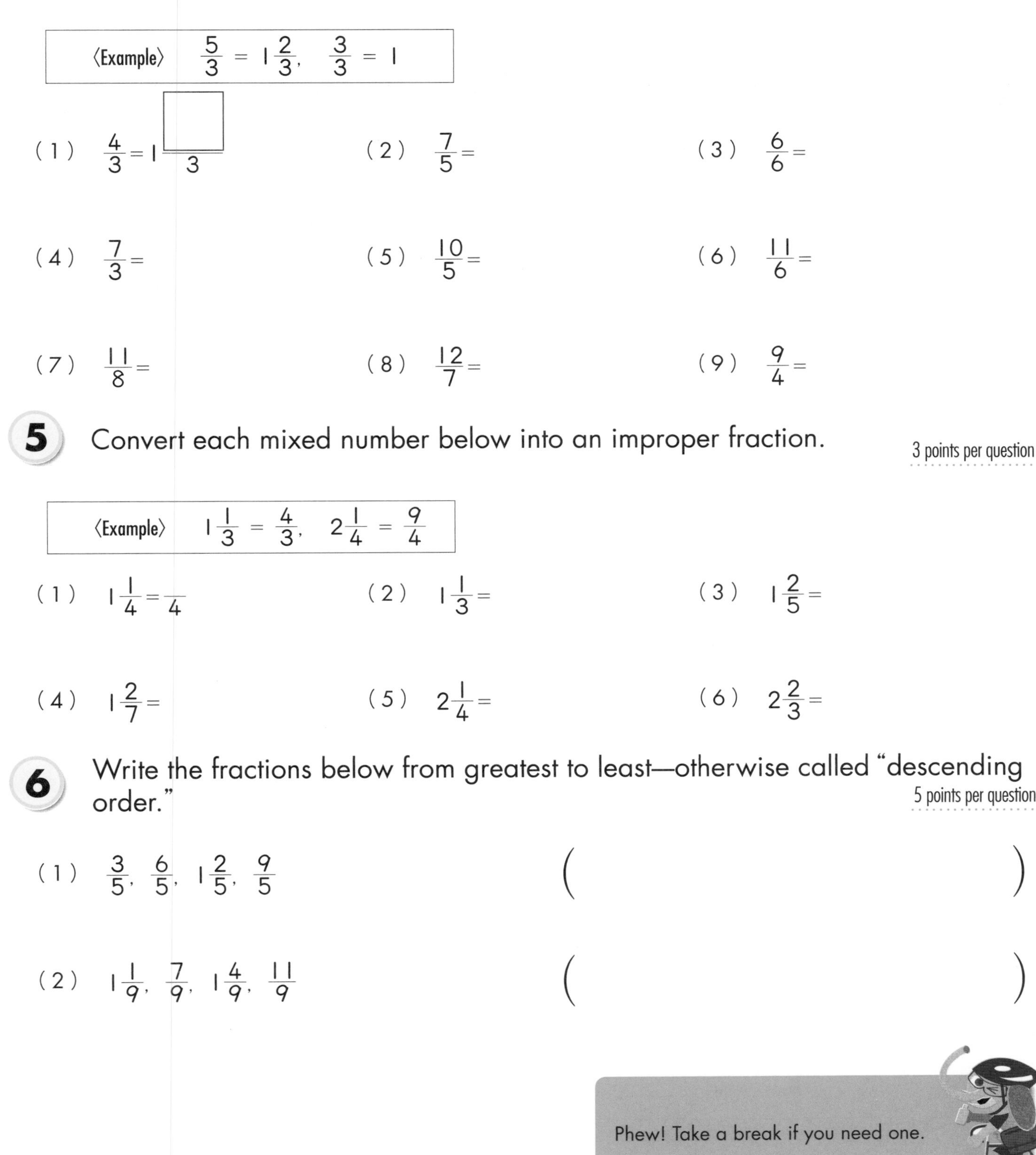

⟨Example⟩ $\frac{5}{3} = 1\frac{2}{3}$, $\frac{3}{3} = 1$

(1) $\frac{4}{3} = 1\frac{\square}{3}$ (2) $\frac{7}{5} =$ (3) $\frac{6}{6} =$

(4) $\frac{7}{3} =$ (5) $\frac{10}{5} =$ (6) $\frac{11}{6} =$

(7) $\frac{11}{8} =$ (8) $\frac{12}{7} =$ (9) $\frac{9}{4} =$

5 Convert each mixed number below into an improper fraction.

3 points per question

⟨Example⟩ $1\frac{1}{3} = \frac{4}{3}$, $2\frac{1}{4} = \frac{9}{4}$

(1) $1\frac{1}{4} = \frac{}{4}$ (2) $1\frac{1}{3} =$ (3) $1\frac{2}{5} =$

(4) $1\frac{2}{7} =$ (5) $2\frac{1}{4} =$ (6) $2\frac{2}{3} =$

6 Write the fractions below from greatest to least—otherwise called "descending order."

5 points per question

(1) $\frac{3}{5}$, $\frac{6}{5}$, $1\frac{2}{5}$, $\frac{9}{5}$ ( )

(2) $1\frac{1}{9}$, $\frac{7}{9}$, $1\frac{4}{9}$, $\frac{11}{9}$ ( )

Phew! Take a break if you need one.

1 Write the appropriate fraction in each box on the number lines below.

2 points per box

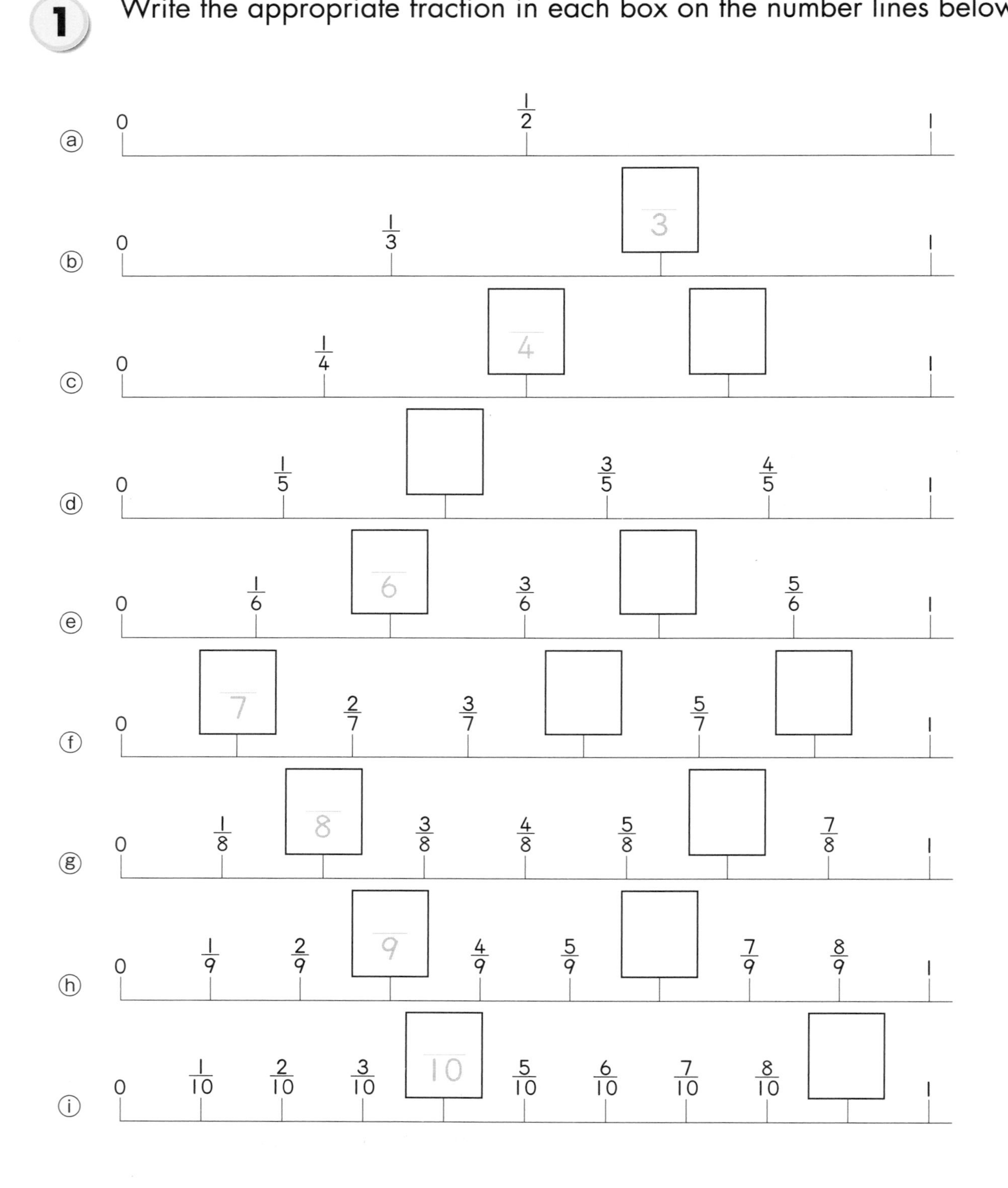

2 Using the number lines on page 22, answer the questions below.

6 points per question

(1) Write all the fractions that are equal to $\frac{1}{2}$. ( )

(2) Write all the fractions that are equal to $\frac{1}{3}$. ( )

3 Write the larger fraction of each pair of fractions in the space provided.

5 points per question

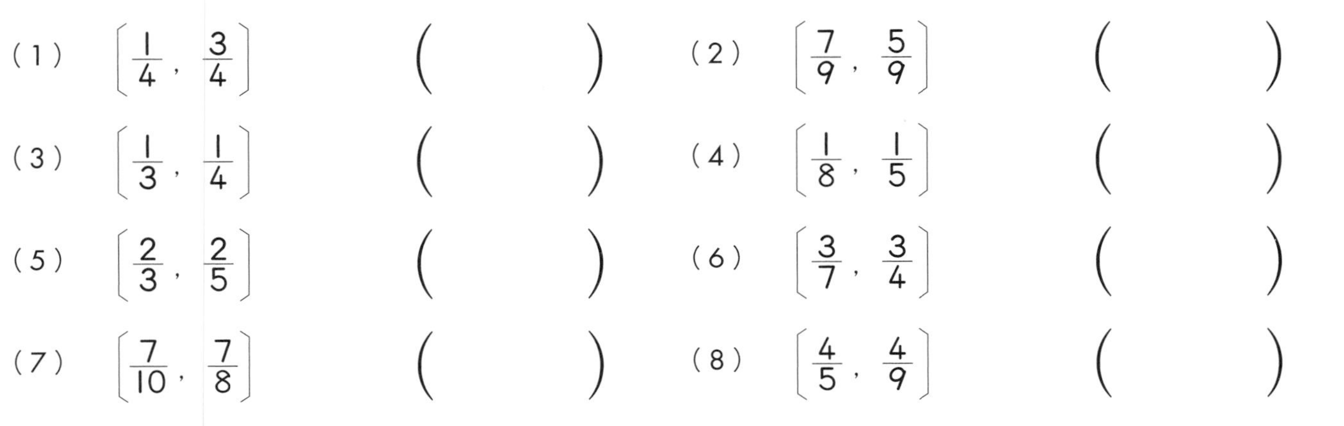

(1) $\left[\frac{1}{4}, \frac{3}{4}\right]$ ( ) (2) $\left[\frac{7}{9}, \frac{5}{9}\right]$ ( )

(3) $\left[\frac{1}{3}, \frac{1}{4}\right]$ ( ) (4) $\left[\frac{1}{8}, \frac{1}{5}\right]$ ( )

(5) $\left[\frac{2}{3}, \frac{2}{5}\right]$ ( ) (6) $\left[\frac{3}{7}, \frac{3}{4}\right]$ ( )

(7) $\left[\frac{7}{10}, \frac{7}{8}\right]$ ( ) (8) $\left[\frac{4}{5}, \frac{4}{9}\right]$ ( )

Don't forget!

When comparing two fractions with the same denominator, the fraction with the larger numerator is the larger number.

When comparing two fractions with the same numerator, the fraction with the smaller denominator is the larger number.

4 Write the fractions below from greatest to least—descending order.

6 points per question

(1) $\frac{4}{9}, \frac{7}{9}, \frac{2}{9}, \frac{8}{9}$ ( )

(2) $\frac{5}{9}, \frac{5}{7}, \frac{5}{8}, \frac{5}{6}$ ( )

(3) $\frac{6}{8}, \frac{6}{5}, \frac{6}{11}, \frac{6}{7}$ ( )

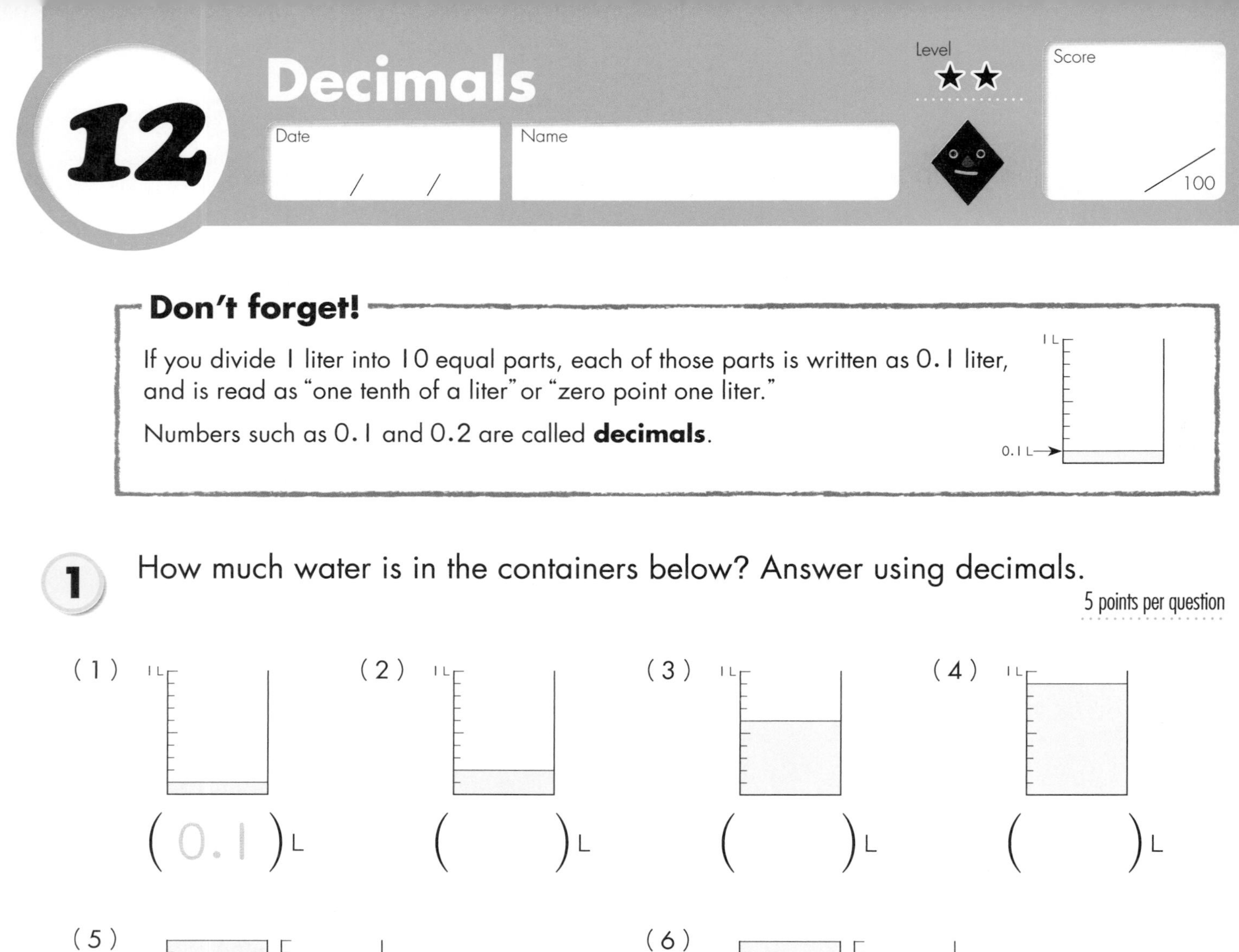

12 Decimals

Level ★★

Date / /

Name

Score /100

Don't forget!

If you divide 1 liter into 10 equal parts, each of those parts is written as 0.1 liter, and is read as "one tenth of a liter" or "zero point one liter."

Numbers such as 0.1 and 0.2 are called **decimals**.

1 How much water is in the containers below? Answer using decimals.

5 points per question

(1) (0.1) L

(2) ( ) L

(3) ( ) L

(4) ( ) L

(5)

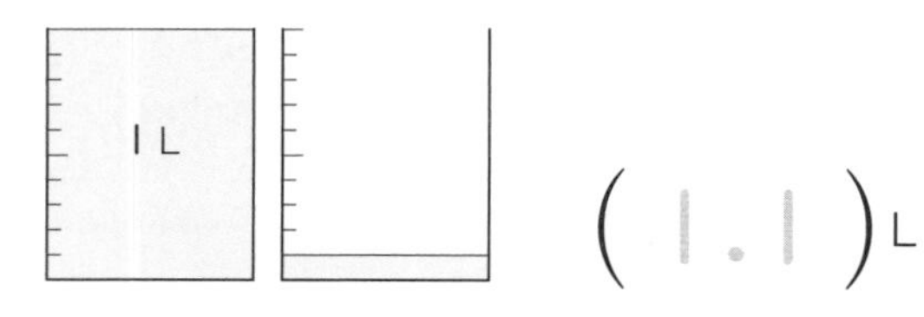

(1.1) L

(6)

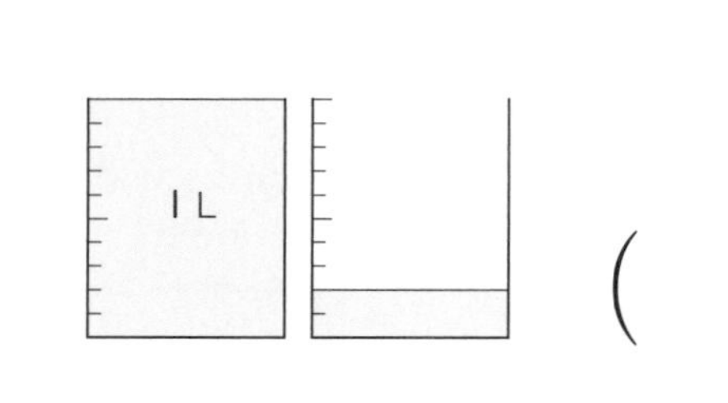

( ) L

(7)

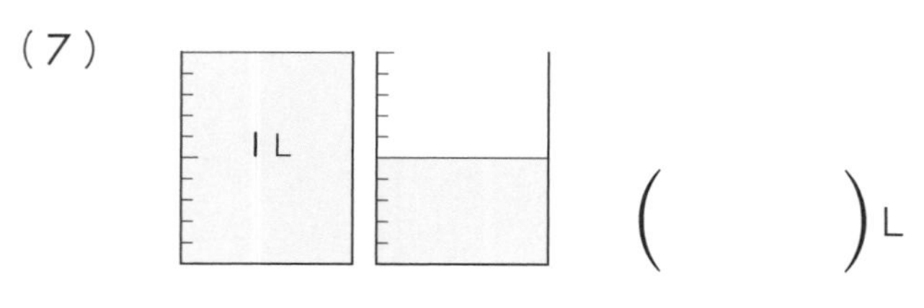

( ) L

(8)

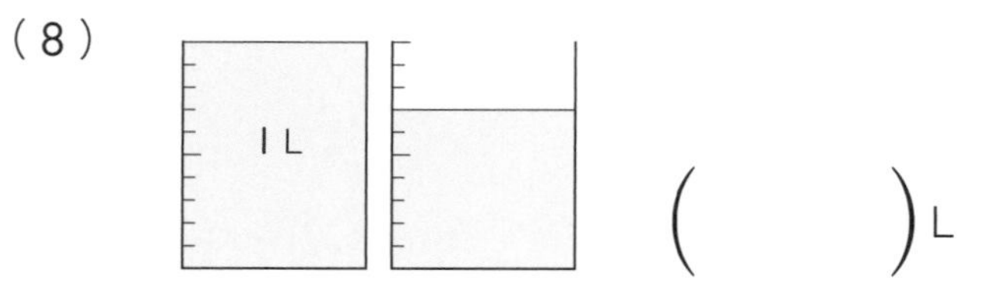

( ) L

(9)

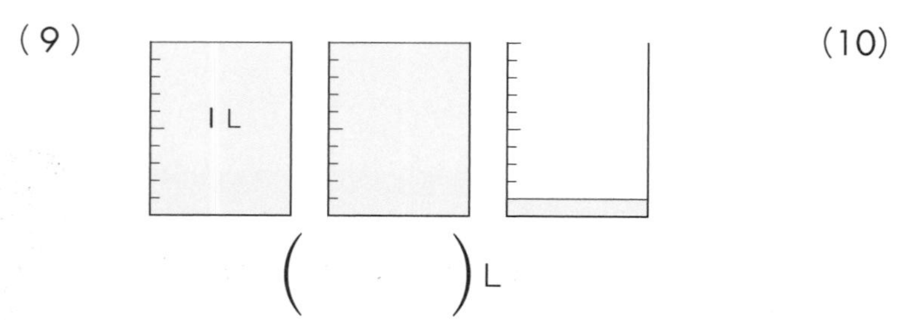

( ) L

(10)

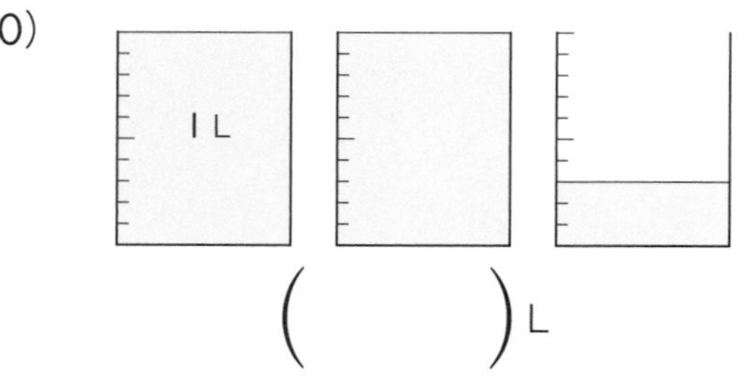

( ) L

2 How much water is in the containers below? Answer using liters first and then write your answer again in milliliters. Remember that 1 L = 1,000 mL.

5 points per question

3 How much water is in the containers below? Answer using liters and then milliliters.

5 points per question

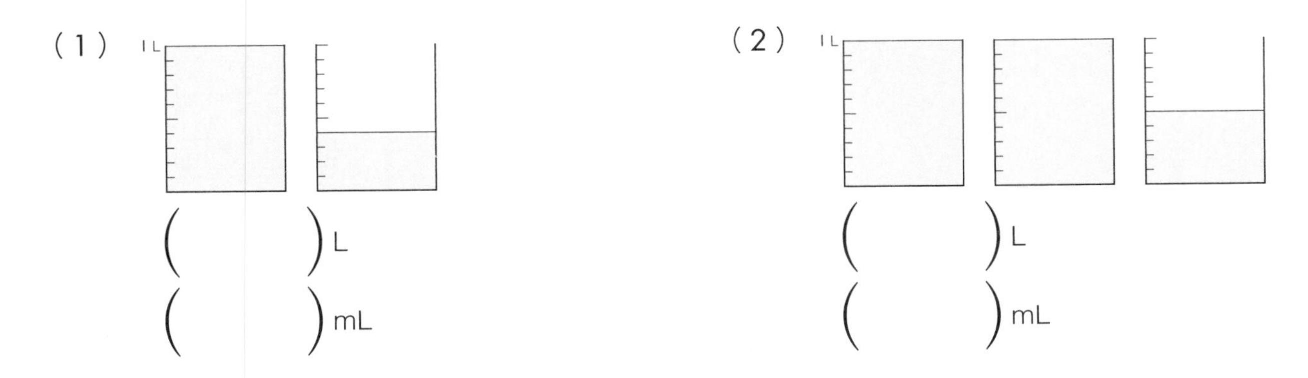

4 Draw the correct amounts of water for each volume given below.

5 points per question

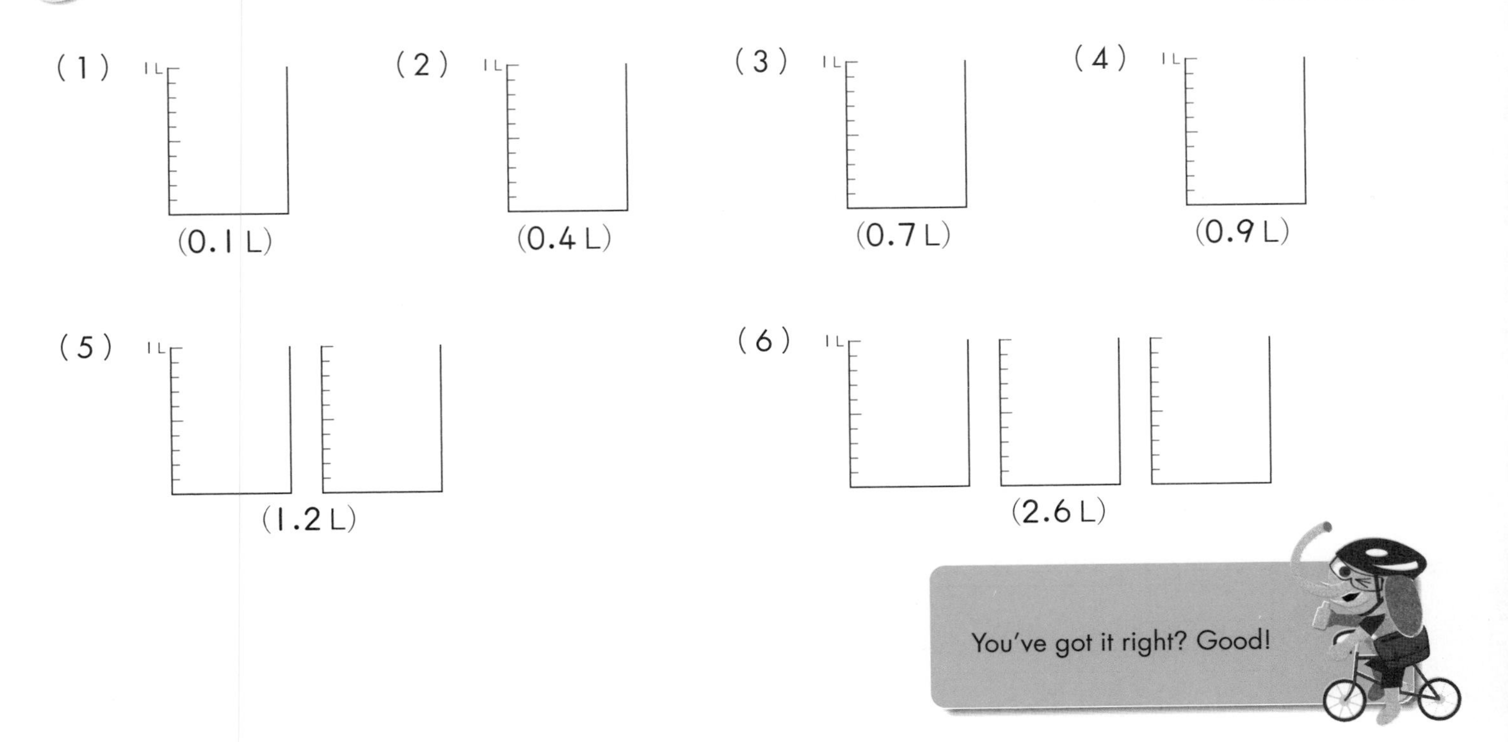

Don't forget!

If you divide 1 centimeter into 10 equal parts, each of those parts is written as 0.1 cm, and is read as "one tenth of a centimeter" or "zero point one centimeter."

0.1cm 1cm

1 How far is it from the left side of the ruler to each box? Answer using decimals.

6 points per question

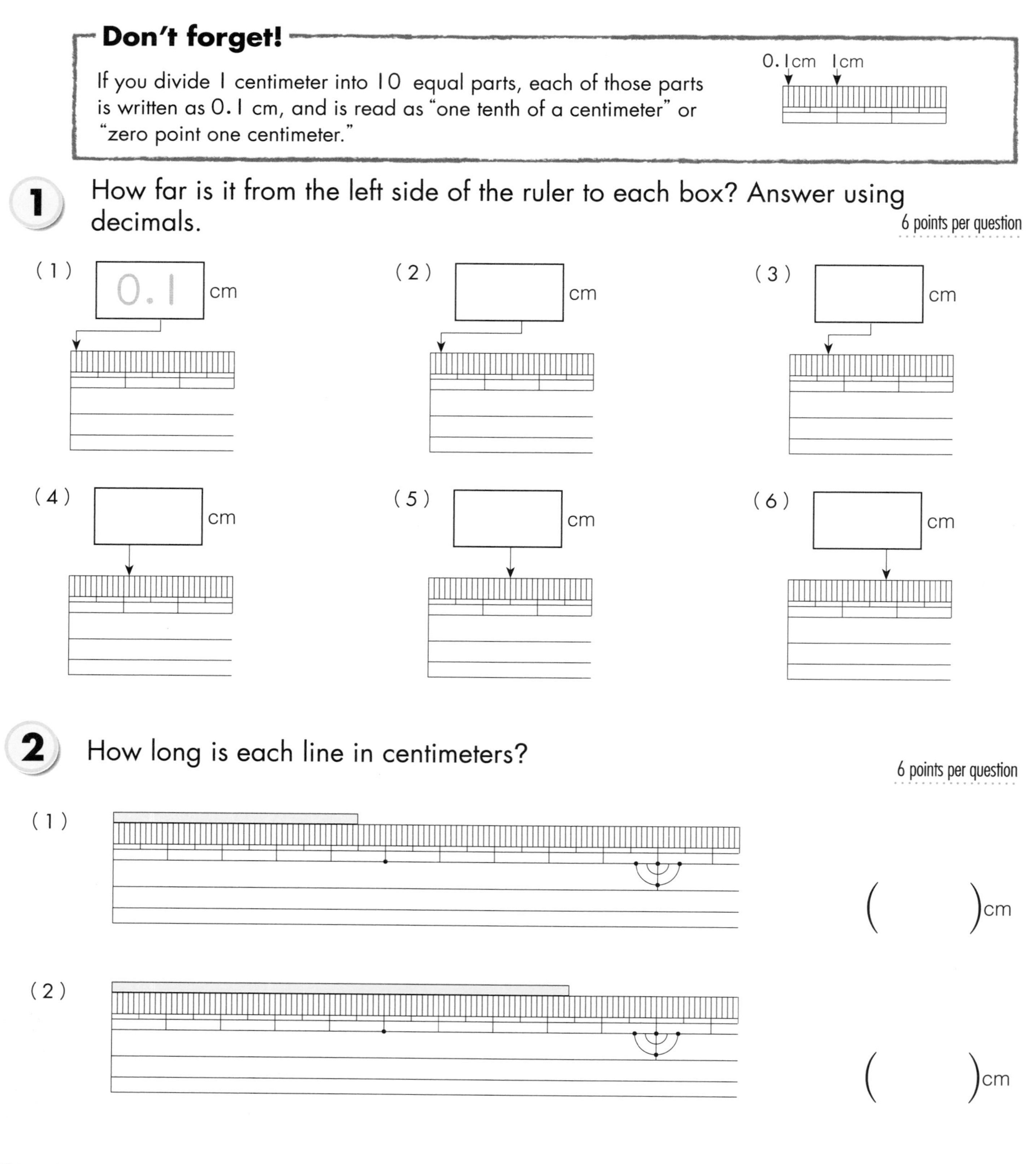

2 How long is each line in centimeters?

6 points per question

(1) ( )cm

(2) ( )cm

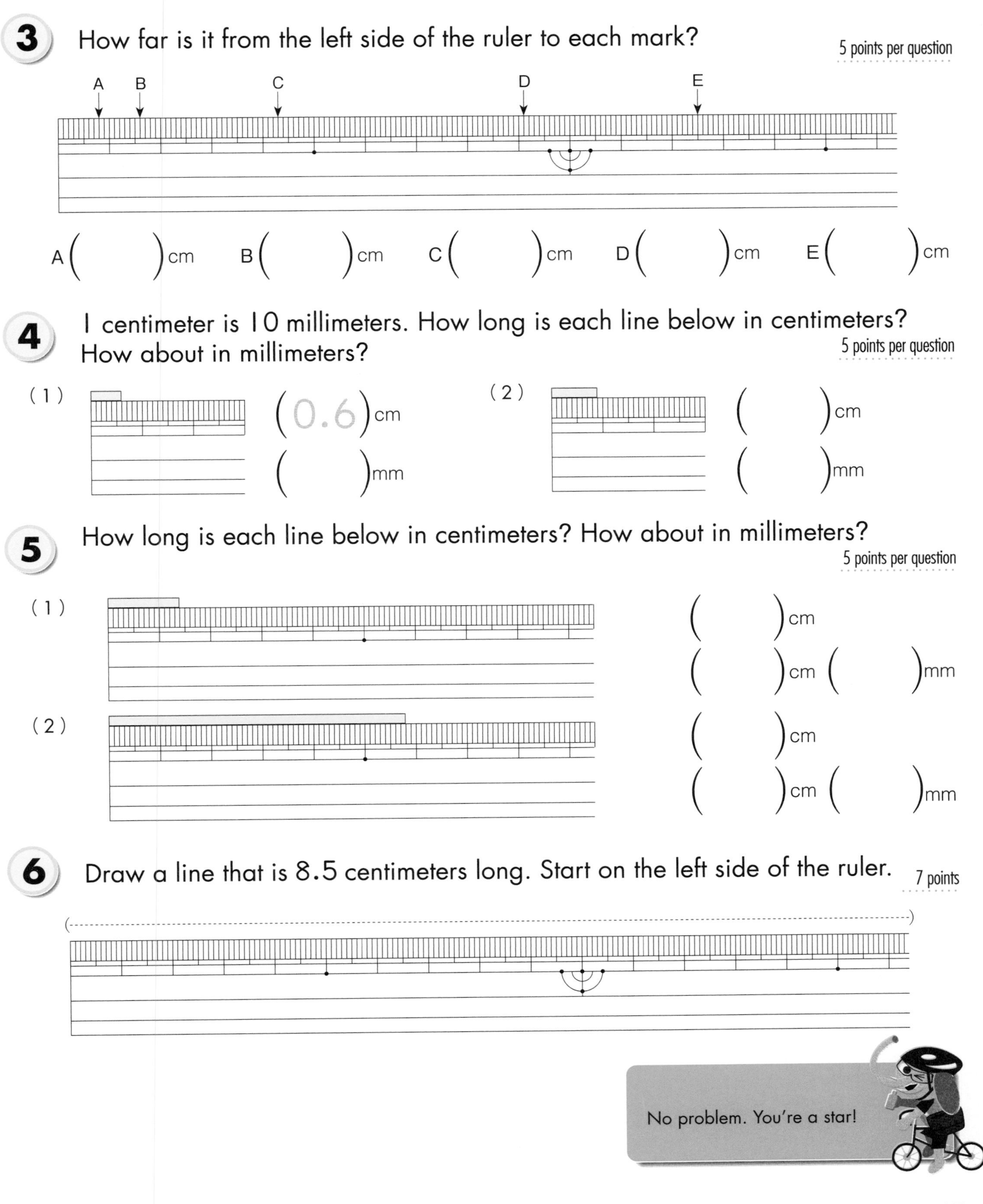

3 How far is it from the left side of the ruler to each mark? 5 points per question

A ( ) cm B ( ) cm C ( ) cm D ( ) cm E ( ) cm

4 1 centimeter is 10 millimeters. How long is each line below in centimeters? How about in millimeters? 5 points per question

(1) (0.6) cm ( ) mm

(2) ( ) cm ( ) mm

5 How long is each line below in centimeters? How about in millimeters? 5 points per question

(1) ( ) cm
( ) cm ( ) mm

(2) ( ) cm
( ) cm ( ) mm

6 Draw a line that is 8.5 centimeters long. Start on the left side of the ruler. 7 points

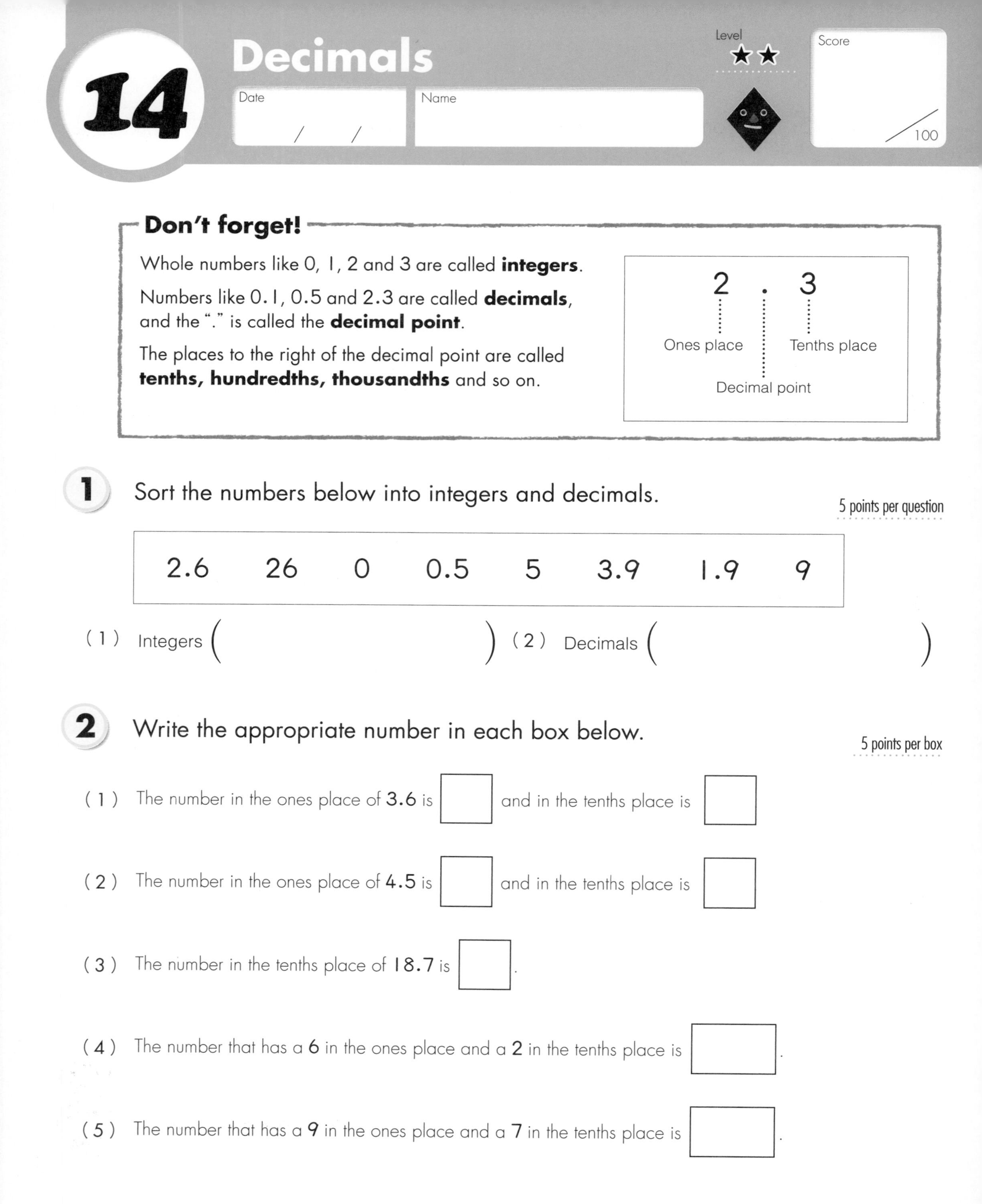

14 Decimals

Level ★★

Date / /

Name

Score /100

Don't forget!

Whole numbers like 0, 1, 2 and 3 are called **integers**.

Numbers like 0.1, 0.5 and 2.3 are called **decimals**, and the "." is called the **decimal point**.

The places to the right of the decimal point are called **tenths, hundredths, thousandths** and so on.

1 Sort the numbers below into integers and decimals.

5 points per question

2.6 26 0 0.5 5 3.9 1.9 9

(1) Integers ( )
(2) Decimals ( )

2 Write the appropriate number in each box below.

5 points per box

(1) The number in the ones place of **3.6** is ☐ and in the tenths place is ☐

(2) The number in the ones place of **4.5** is ☐ and in the tenths place is ☐

(3) The number in the tenths place of **18.7** is ☐.

(4) The number that has a **6** in the ones place and a **2** in the tenths place is ☐.

(5) The number that has a **9** in the ones place and a **7** in the tenths place is ☐.

3 Using the figure pictured here, write the appropriate number in each box below.

5 points per question

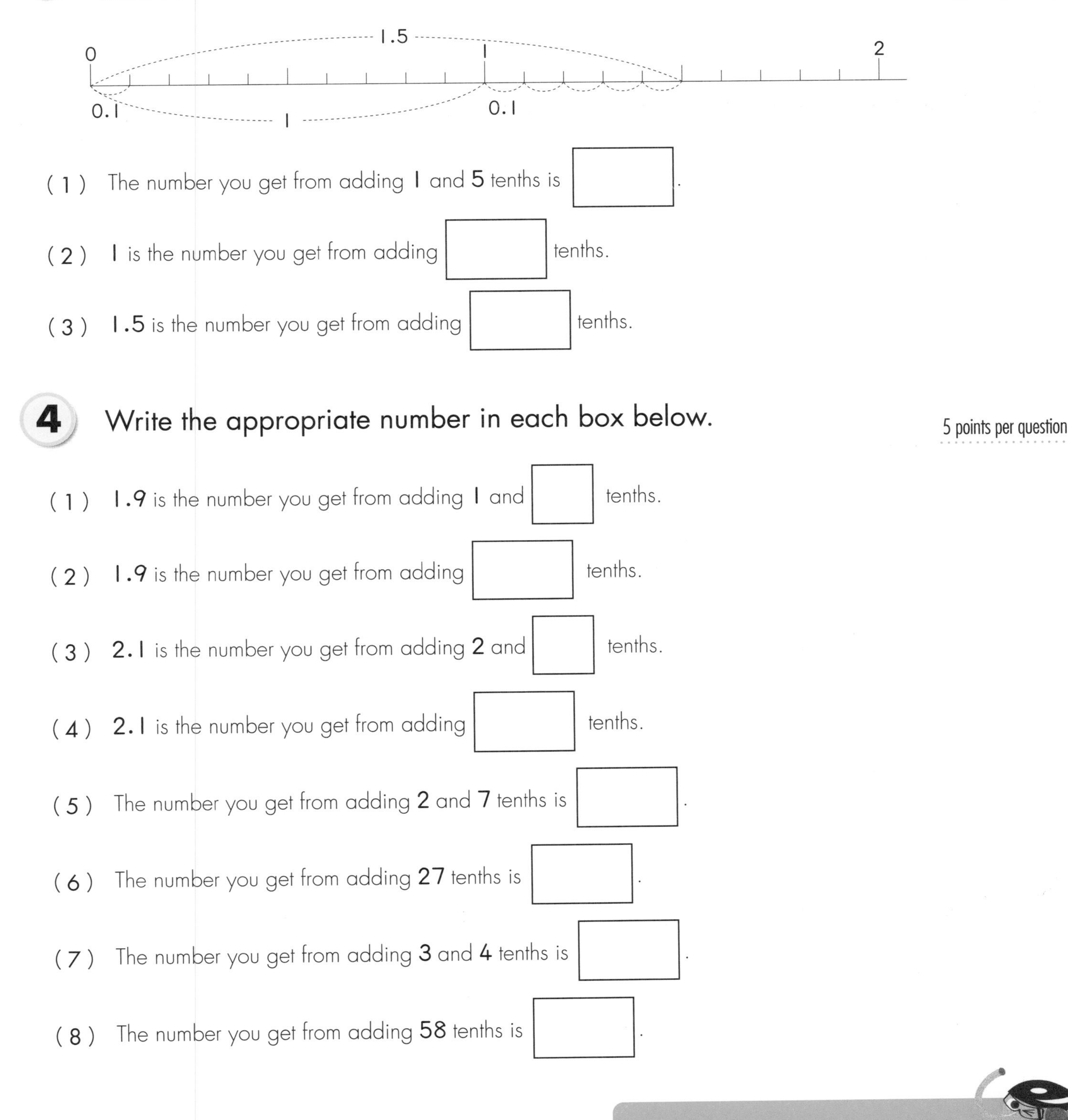

(1) The number you get from adding 1 and 5 tenths is ☐.

(2) 1 is the number you get from adding ☐ tenths.

(3) 1.5 is the number you get from adding ☐ tenths.

4 Write the appropriate number in each box below.

5 points per question

(1) 1.9 is the number you get from adding 1 and ☐ tenths.

(2) 1.9 is the number you get from adding ☐ tenths.

(3) 2.1 is the number you get from adding 2 and ☐ tenths.

(4) 2.1 is the number you get from adding ☐ tenths.

(5) The number you get from adding 2 and 7 tenths is ☐.

(6) The number you get from adding 27 tenths is ☐.

(7) The number you get from adding 3 and 4 tenths is ☐.

(8) The number you get from adding 58 tenths is ☐.

Decimals aren't so bad. You've got it!

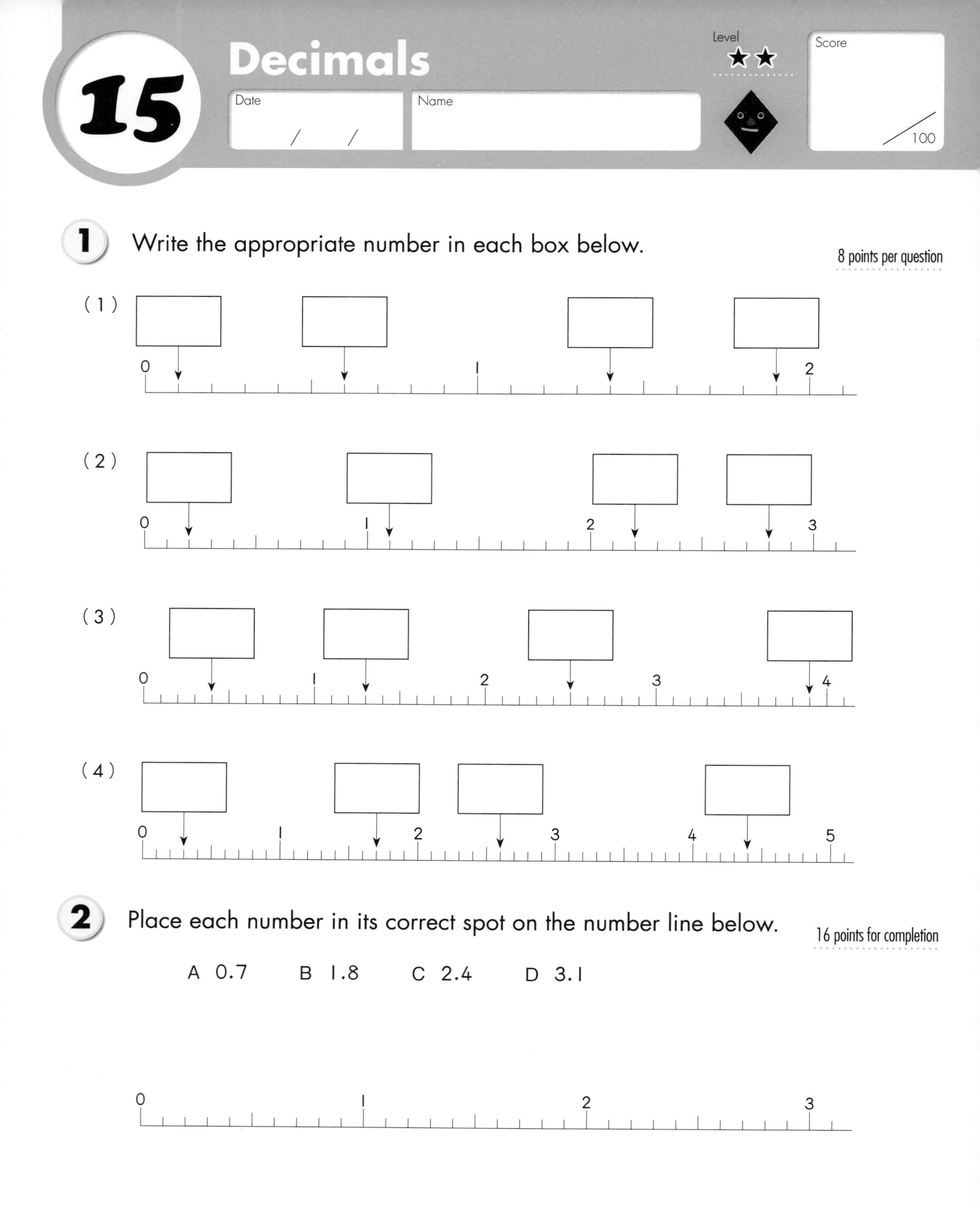

15 Decimals

Level ★★

Date / / Name

Score /100

1 Write the appropriate number in each box below.

8 points per question

(1)

(2)

(3)

(4)

2 Place each number in its correct spot on the number line below.

16 points for completion

A 0.7 B 1.8 C 2.4 D 3.1

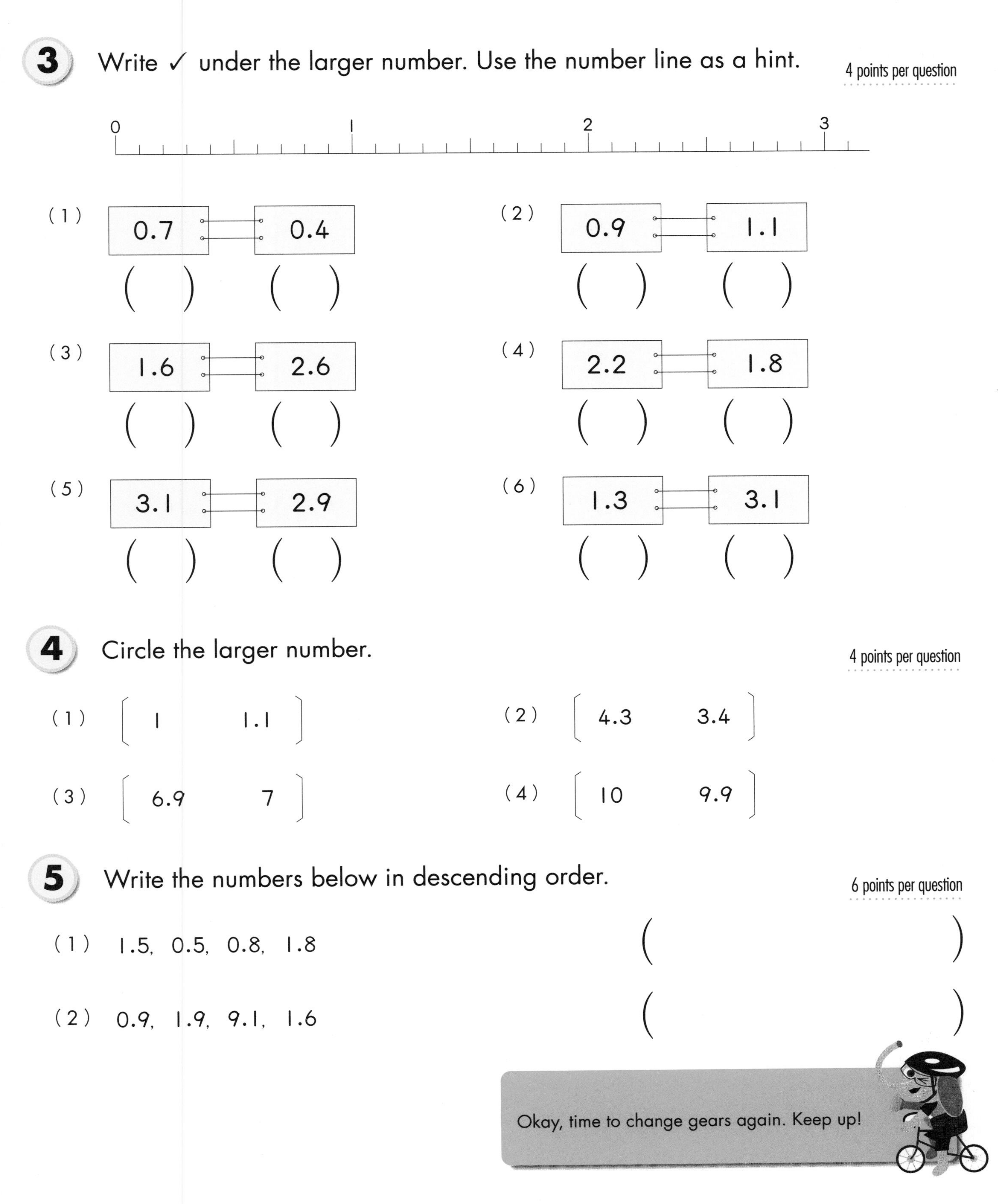

3 Write ✓ under the larger number. Use the number line as a hint.

4 points per question

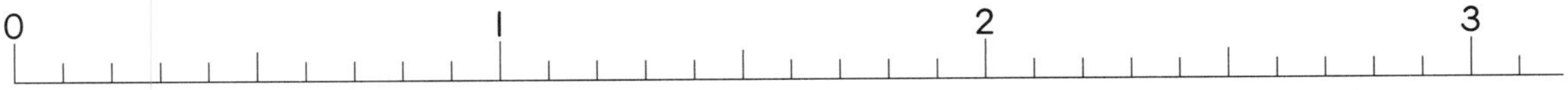

(1) 0.7 ( ) 0.4 ( )

(2) 0.9 ( ) 1.1 ( )

(3) 1.6 ( ) 2.6 ( )

(4) 2.2 ( ) 1.8 ( )

(5) 3.1 ( ) 2.9 ( )

(6) 1.3 ( ) 3.1 ( )

4 Circle the larger number.

4 points per question

(1) [1 1.1]

(2) [4.3 3.4]

(3) [6.9 7]

(4) [10 9.9]

5 Write the numbers below in descending order.

6 points per question

(1) 1.5, 0.5, 0.8, 1.8 ( )

(2) 0.9, 1.9, 9.1, 1.6 ( )

Okay, time to change gears again. Keep up!

Don't forget!

The volume of a cube with sides of 1 inch is called 1 cubic inch, and is written 1 in.^{3}

1 in.
1 in.
1 in.
1 in.^{3}

1 The following shapes were made by cubes with 1-inch sides. Calculate the volume of each shape below.

4 points per question

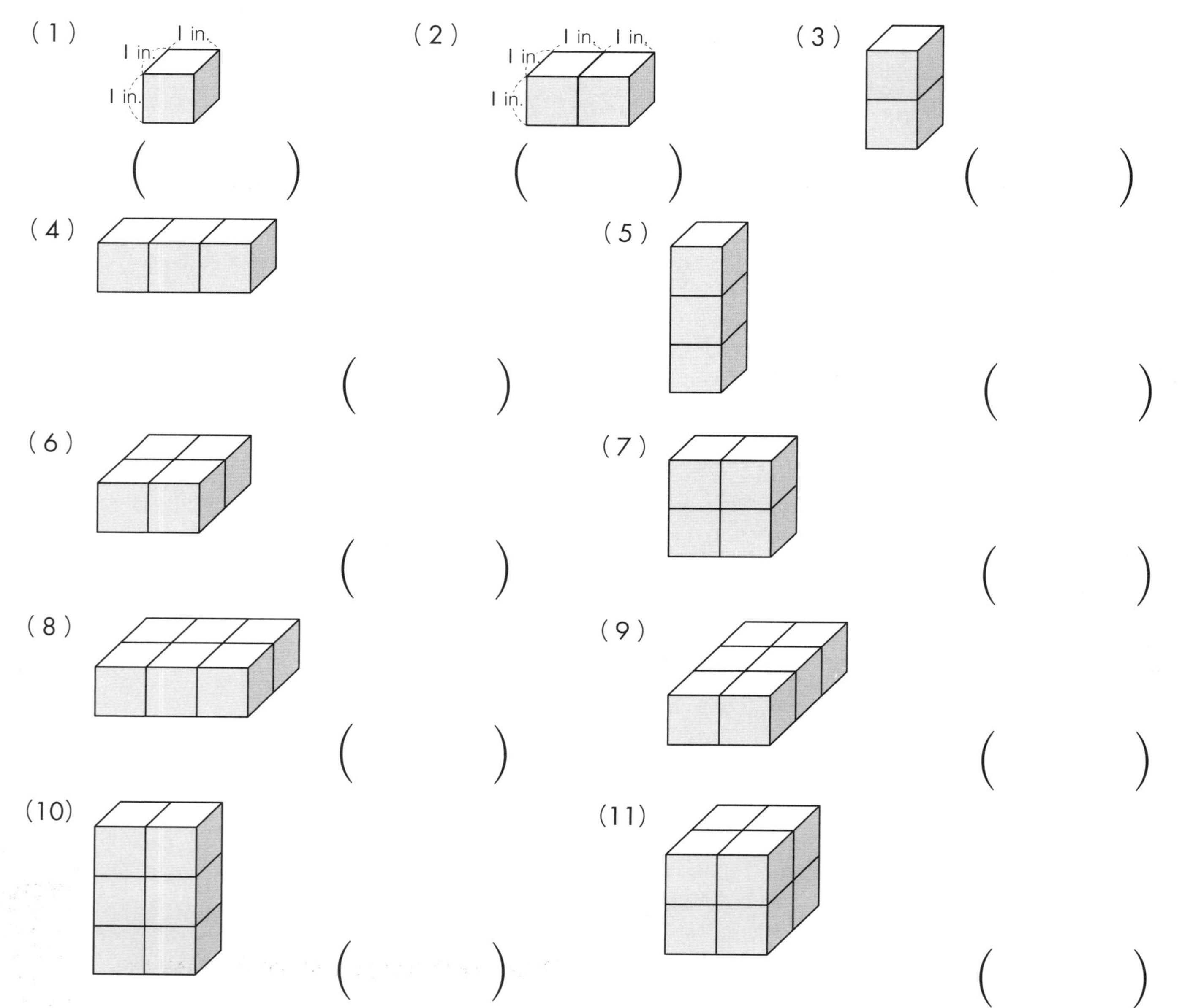

2 Calculate the volume of the following rectangular, solid shapes—also called prisms. Answer in cubic inches.

7 points per question

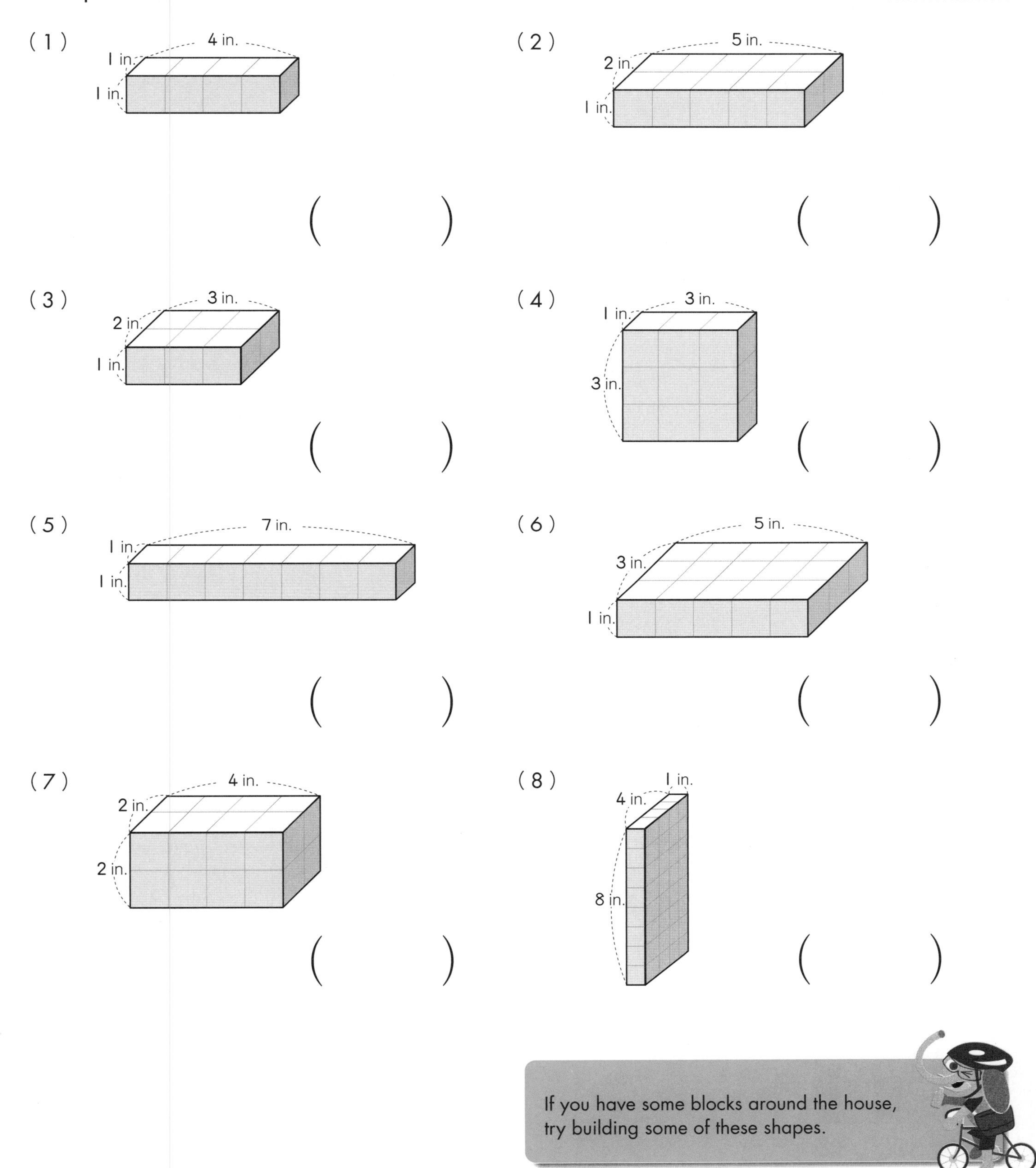

17 Volume

Level ★★

Date / /

Name

Score /100

Don't forget!

The volume of a cube with sides of 1 centimeter is called 1 cubic centimeter, and is written 1 cm^3.

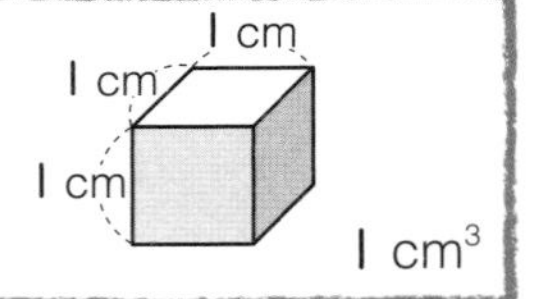

1 The following shapes were made by cubes with 1-centimeter sides. Calculate the volume of each shape below.

4 points per question

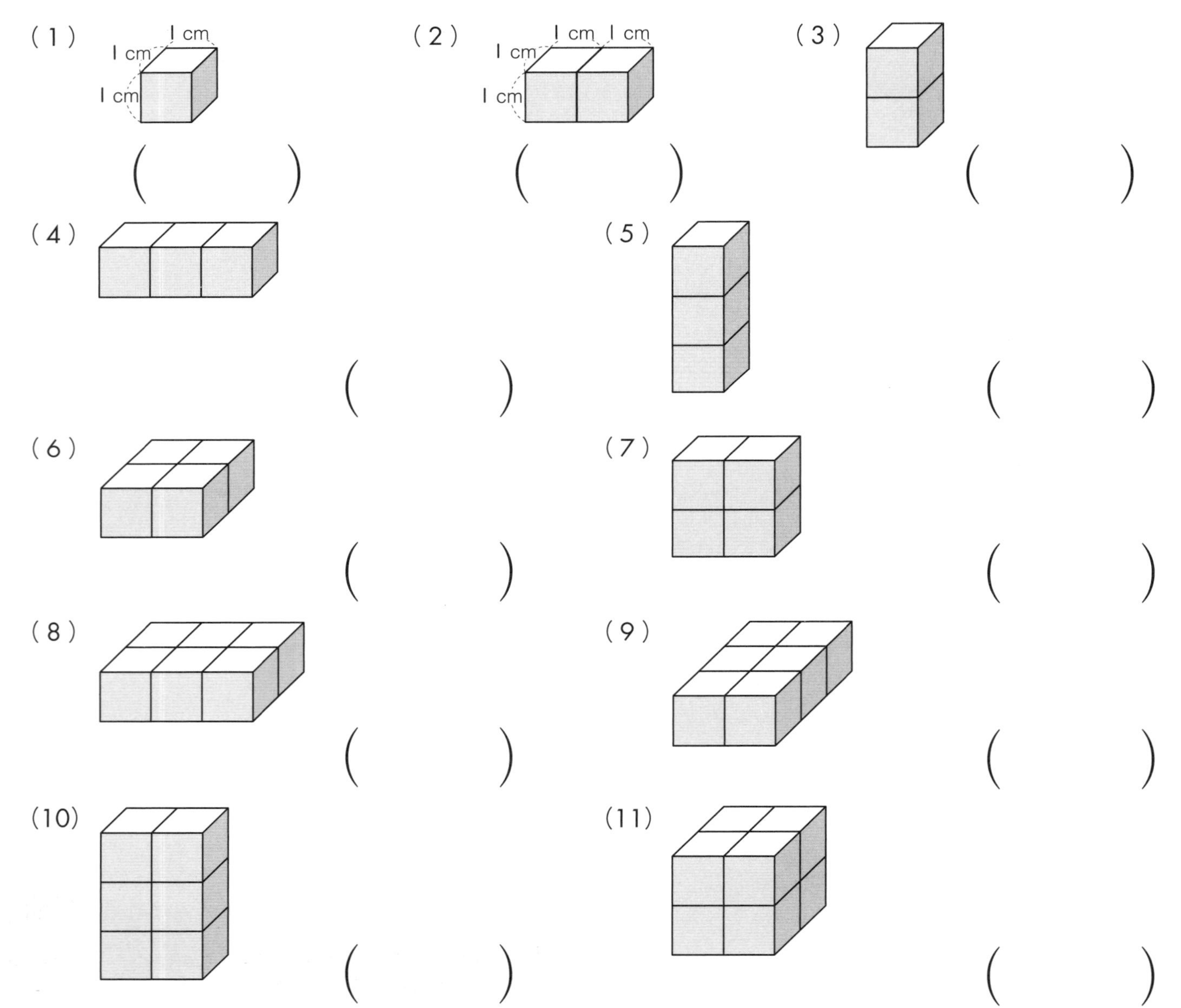

2 What is the volume of the following rectangular prisms? Answer in cubic centimeters.

7 points per question

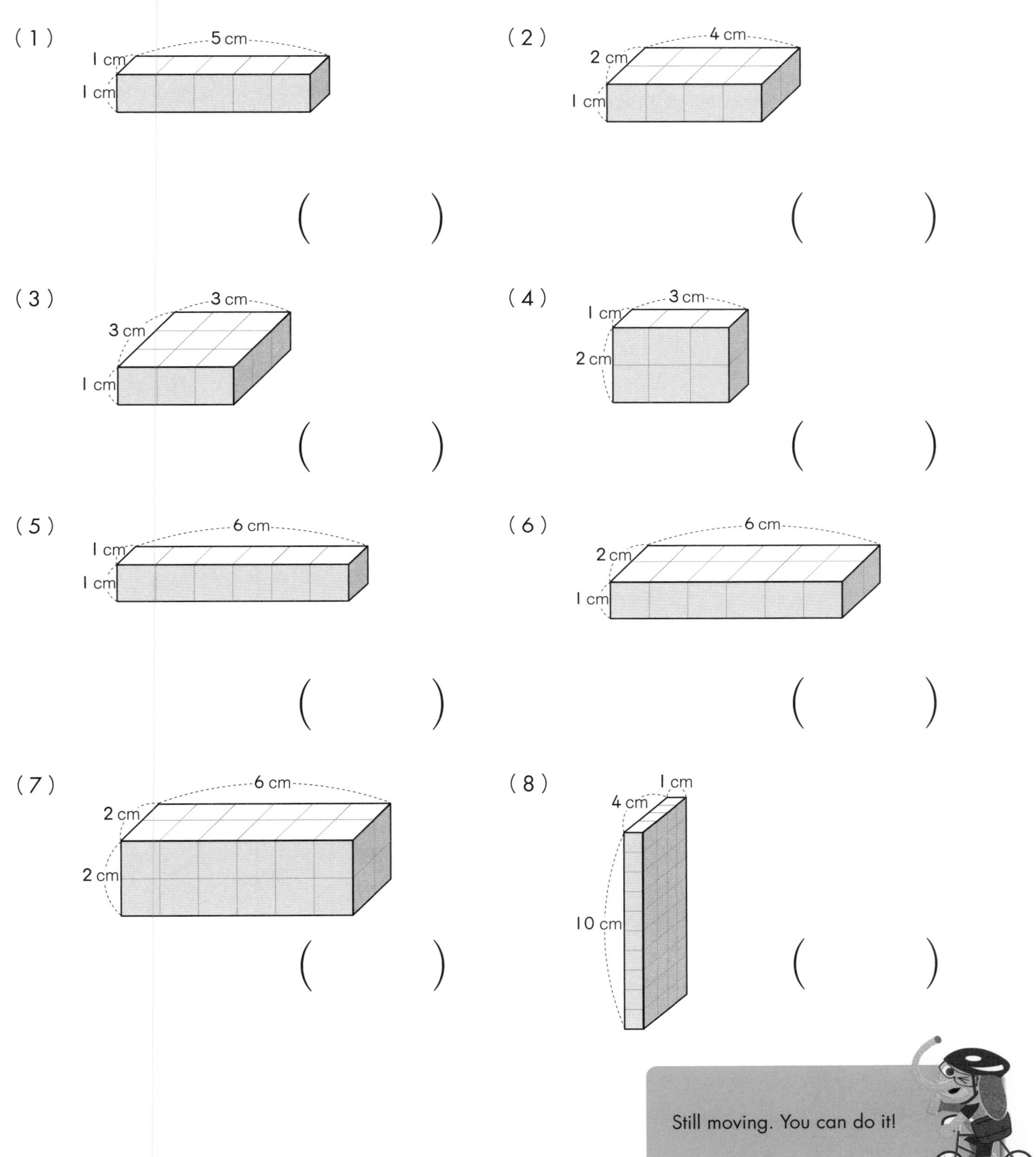

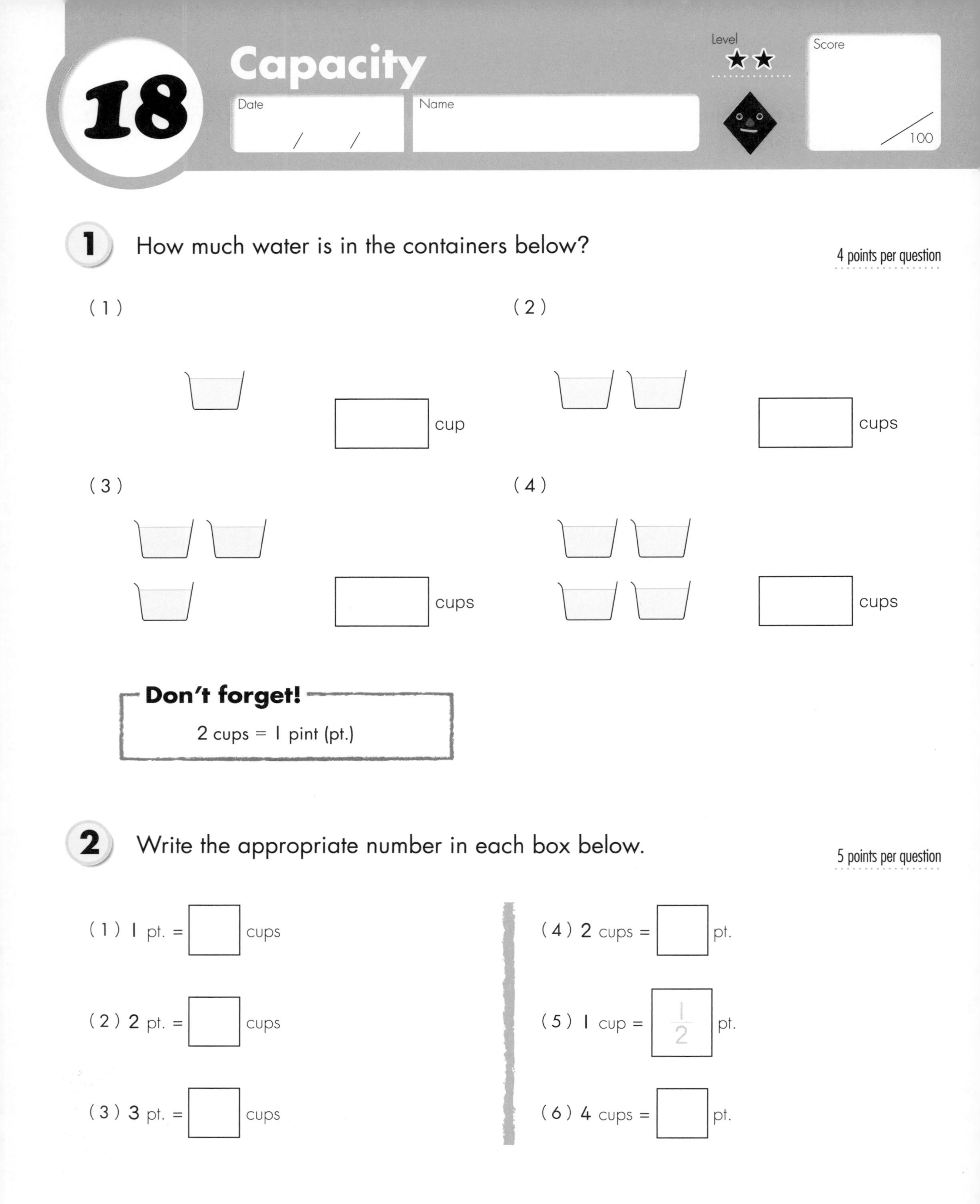

18 Capacity

Date / / Name

Level ★★

Score /100

1 How much water is in the containers below?

4 points per question

(1) ☐ cup

(2) ☐ cups

(3) ☐ cups

(4) ☐ cups

Don't forget!

2 cups = 1 pint (pt.)

2 Write the appropriate number in each box below.

5 points per question

(1) 1 pt. = ☐ cups

(2) 2 pt. = ☐ cups

(3) 3 pt. = ☐ cups

(4) 2 cups = ☐ pt.

(5) 1 cup = $\frac{1}{2}$ pt.

(6) 4 cups = ☐ pt.

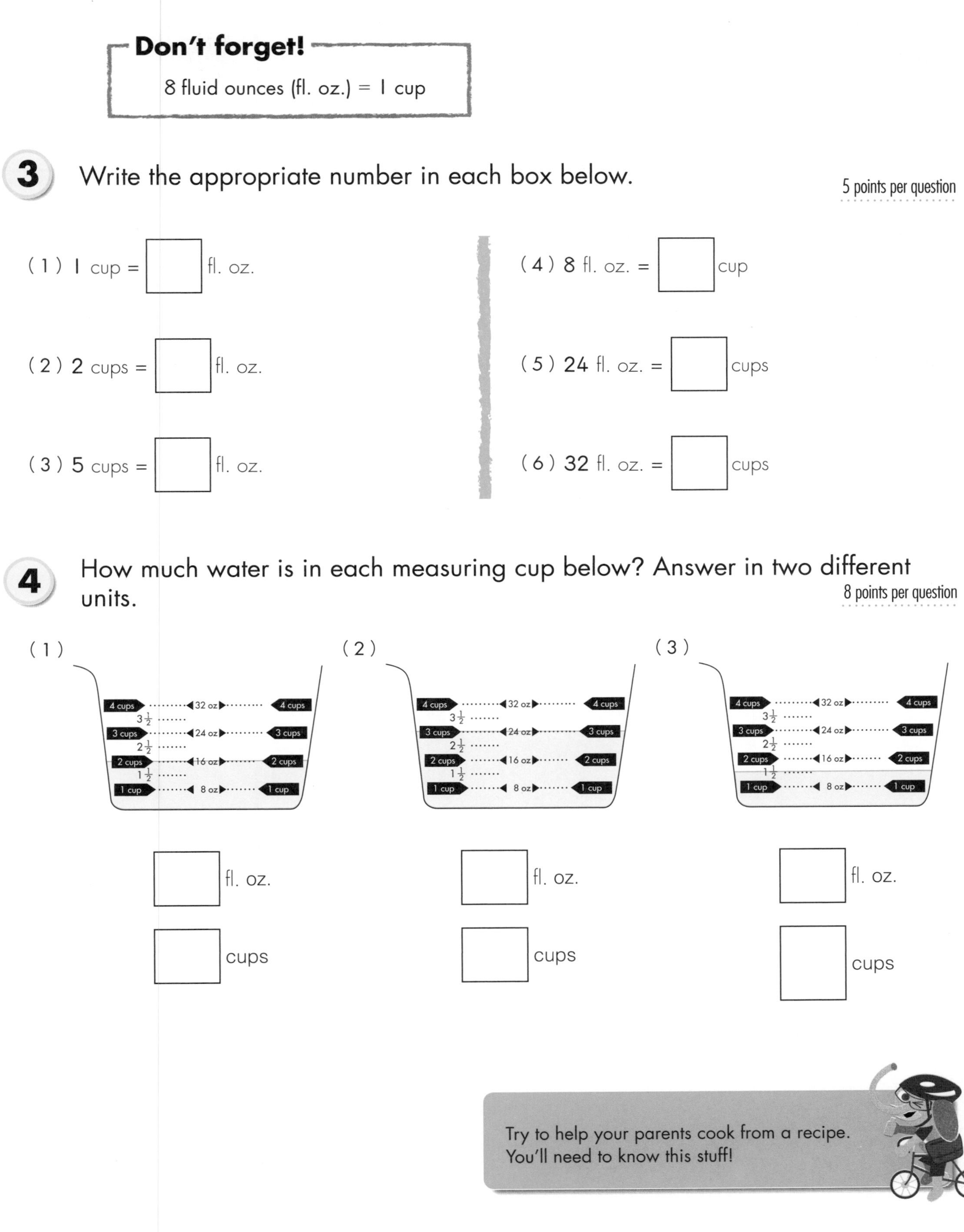
Don't forget!
8 fluid ounces (fl. oz.) = 1 cup
3 Write the appropriate number in each box below.
5 points per question
(1) 1 cup = ☐ fl. oz.
(2) 2 cups = ☐ fl. oz.
(3) 5 cups = ☐ fl. oz.
(4) 8 fl. oz. = ☐ cup
(5) 24 fl. oz. = ☐ cups
(6) 32 fl. oz. = ☐ cups
4 How much water is in each measuring cup below? Answer in two different units.
8 points per question
(1)
(2)
(3)
4 cups
32 oz
3 1/2
3 cups
24 oz
2 1/2
2 cups
16 oz
1 1/2
1 cup
8 oz
☐ fl. oz.
☐ cups
Try to help your parents cook from a recipe.
You'll need to know this stuff!

19 Area

Level ★★

Date / /

Name

Score /100

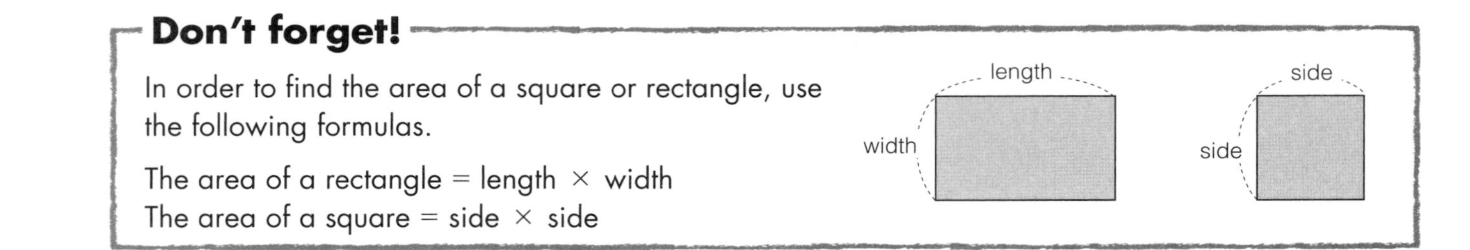

Don't forget!

In order to find the area of a square or rectangle, use the following formulas.

The area of a rectangle = length × width
The area of a square = side × side

1 What is the area of each shape below? Answer in square inches and use the formulas from above.

6 points per question

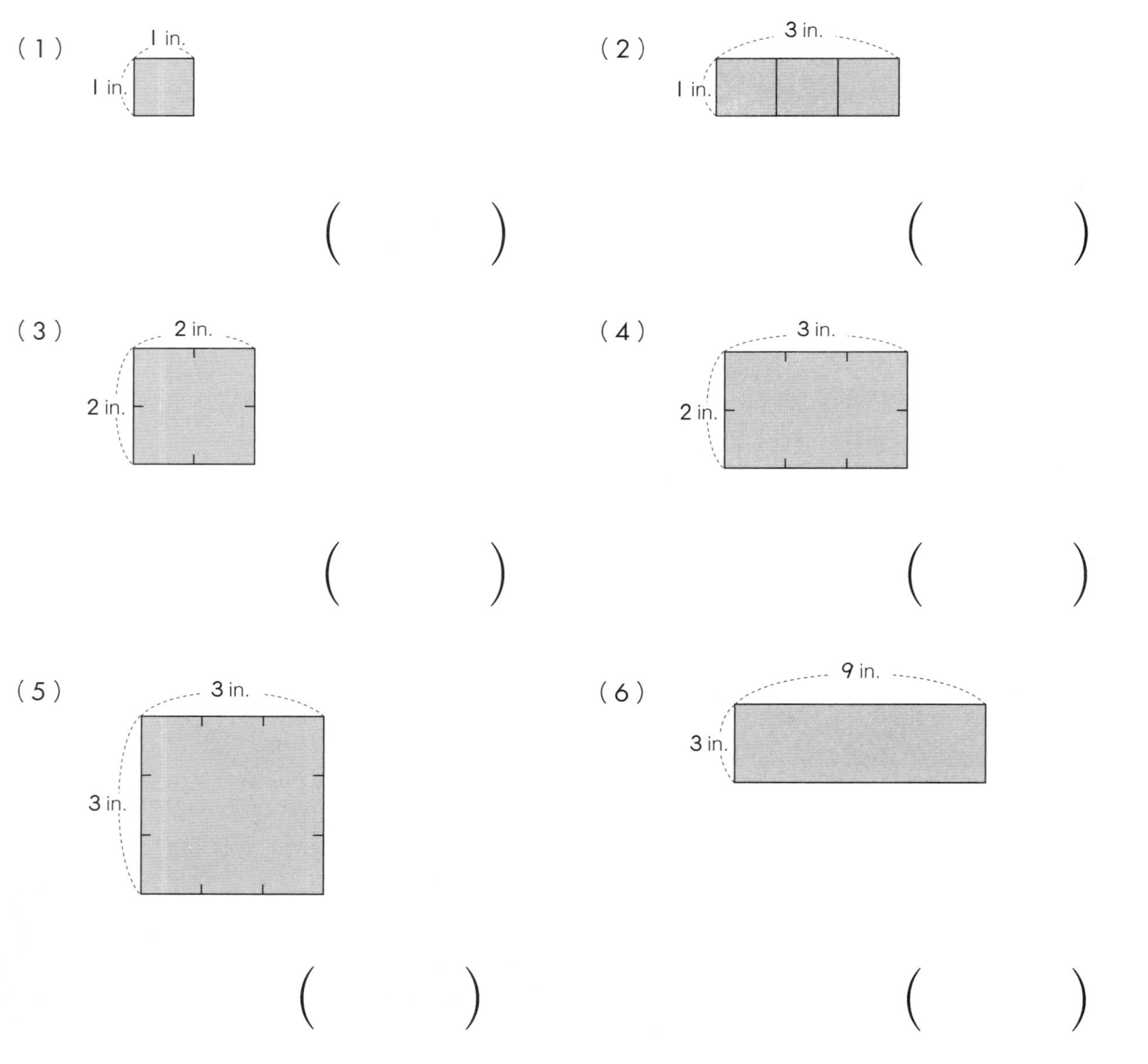

( )

( )

( )

( )

( )

( )

Don't forget!

A square with sides of 1 foot (ft.) has an area of one square foot, written as 1 $ft.^2$

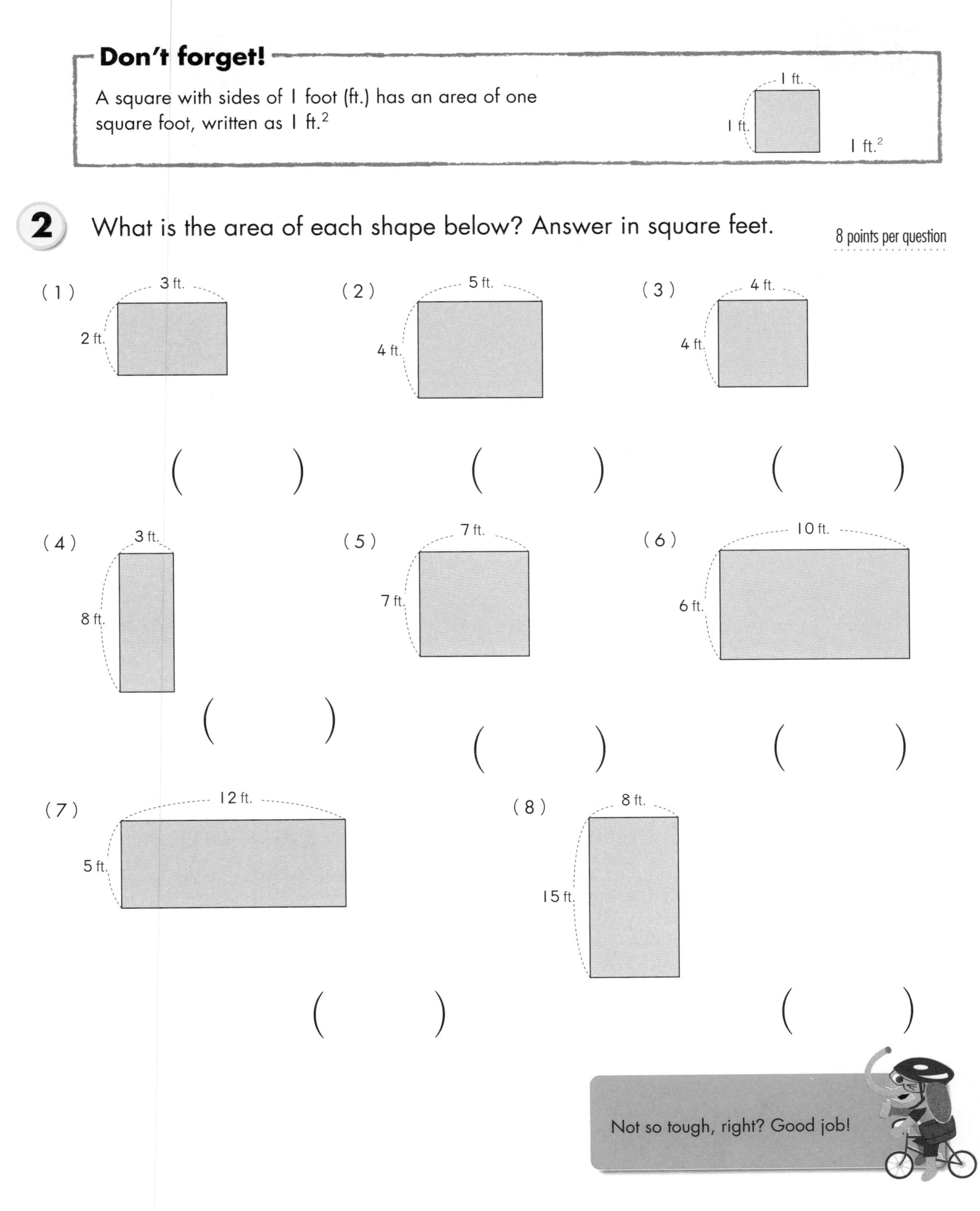

2 What is the area of each shape below? Answer in square feet.

8 points per question

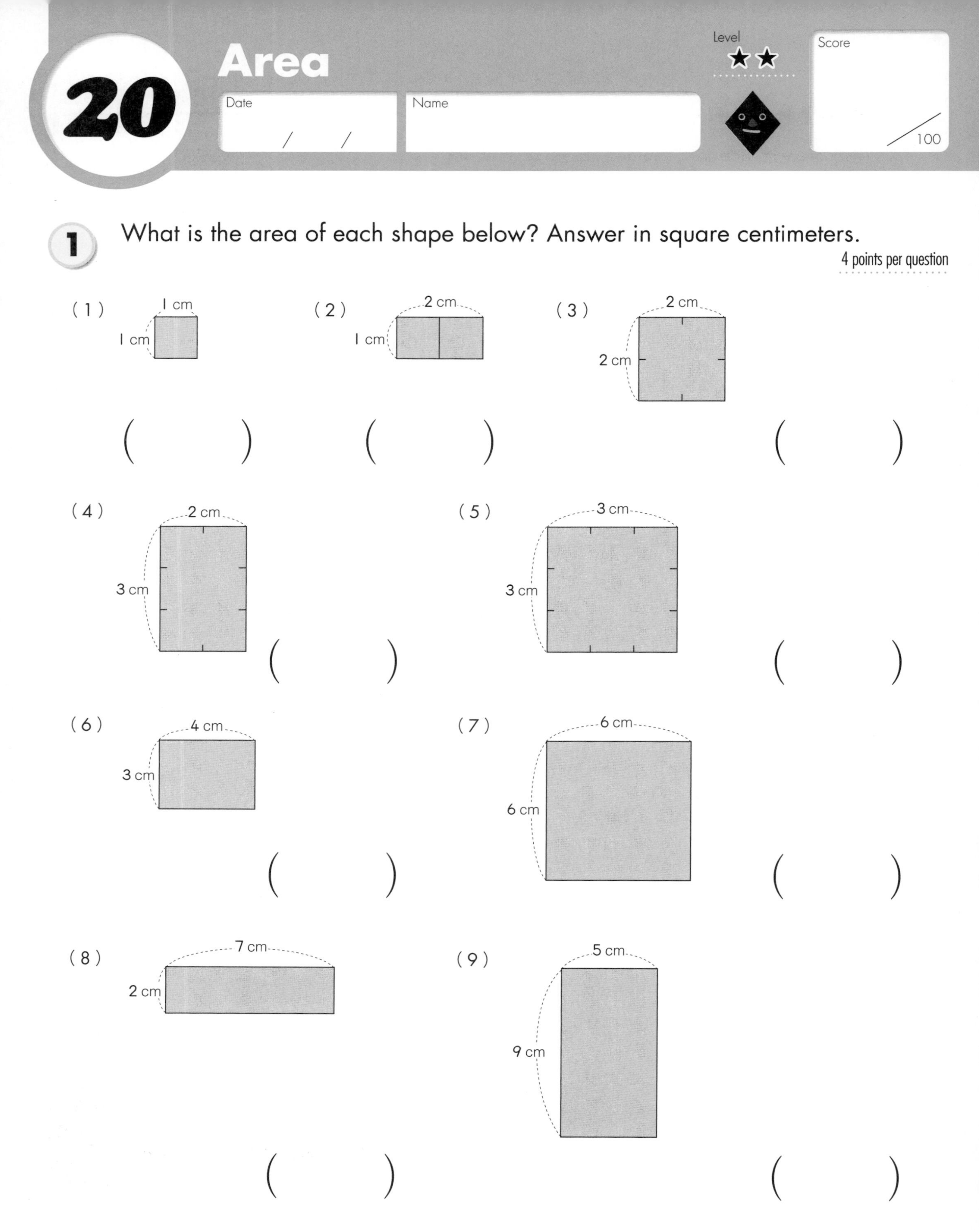
20
Area
Level
Date
/ /
Name
Score
100
1
What is the area of each shape below? Answer in square centimeters.
4 points per question
(1)
1 cm
1 cm
(2)
2 cm
1 cm
(3)
2 cm
2 cm
(4)
2 cm
3 cm
(5)
3 cm
3 cm
(6)
4 cm
3 cm
(7)
6 cm
6 cm
(8)
7 cm
2 cm
(9)
5 cm
9 cm

Don't forget!

A square with sides of 1 meter (m) has an area of one square meter, written as 1 m^2.

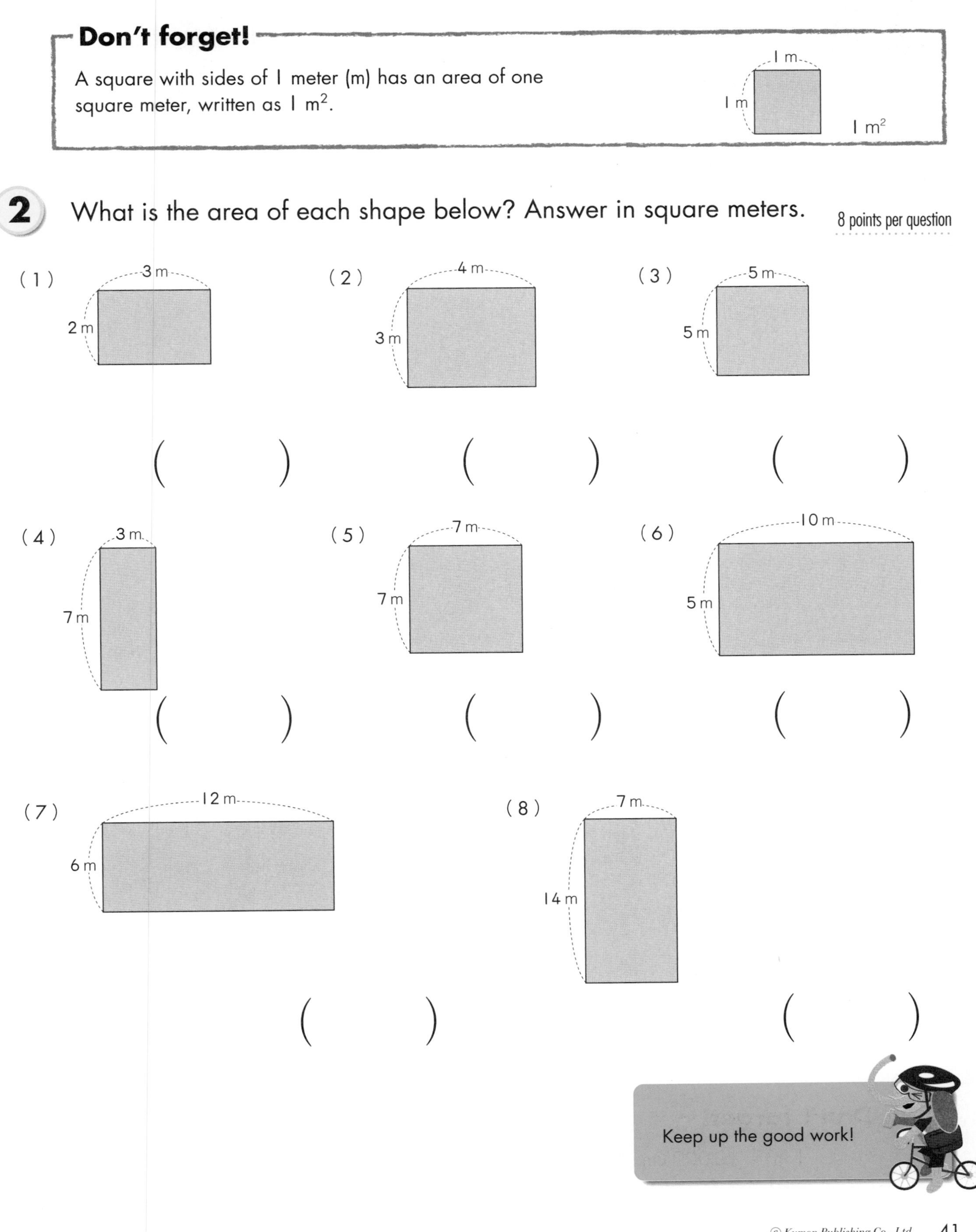

2 What is the area of each shape below? Answer in square meters. 8 points per question

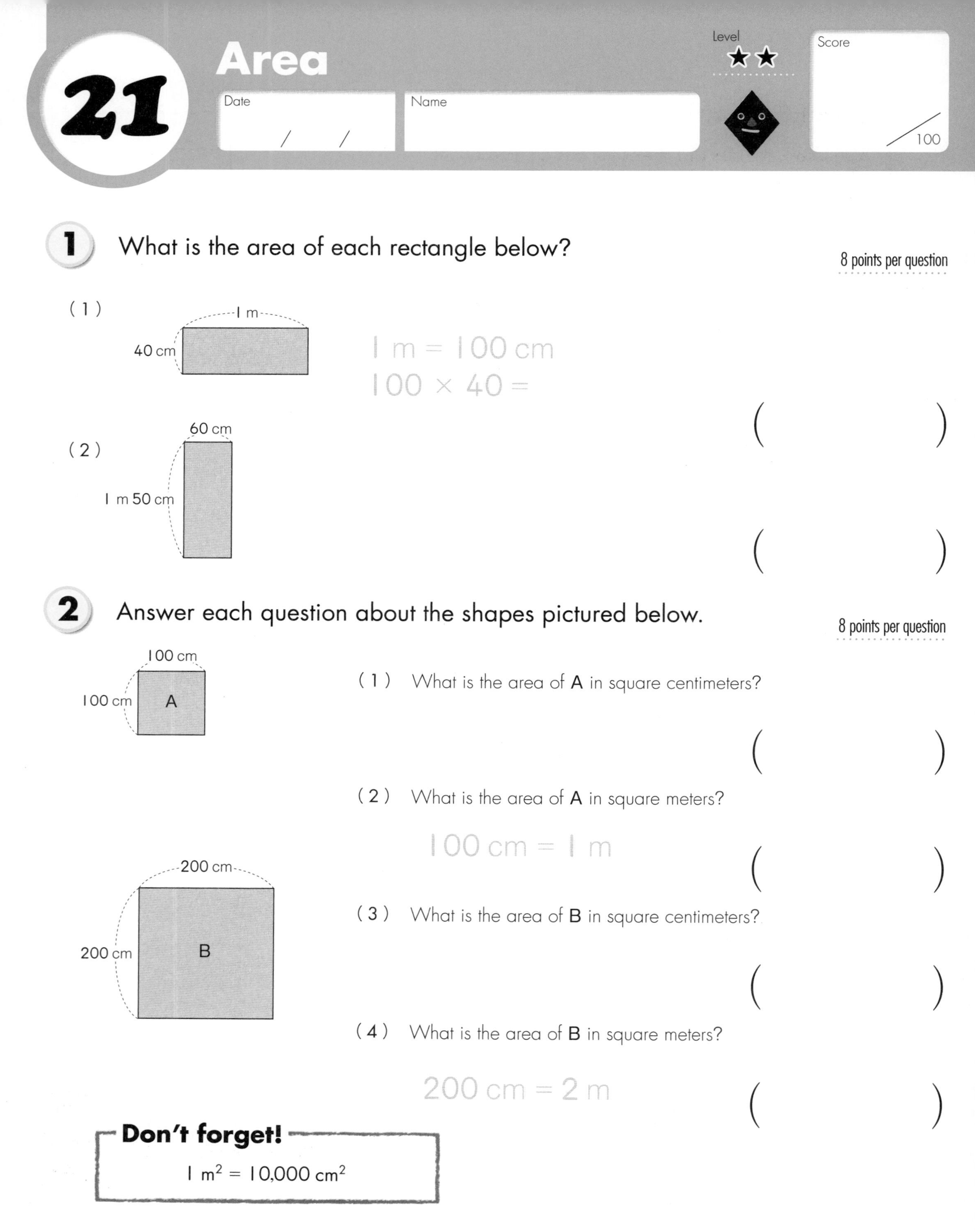

21 Area

Date / /

Name

Level ★★

Score /100

1 What is the area of each rectangle below?

8 points per question

(1)

1 m = 100 cm

100 × 40 =

( )

(2)

( )

2 Answer each question about the shapes pictured below.

8 points per question

(1) What is the area of A in square centimeters?

( )

(2) What is the area of A in square meters?

100 cm = 1 m

( )

(3) What is the area of B in square centimeters?

( )

(4) What is the area of B in square meters?

200 cm = 2 m

( )

Don't forget!

1 m^2 = 10,000 cm^2

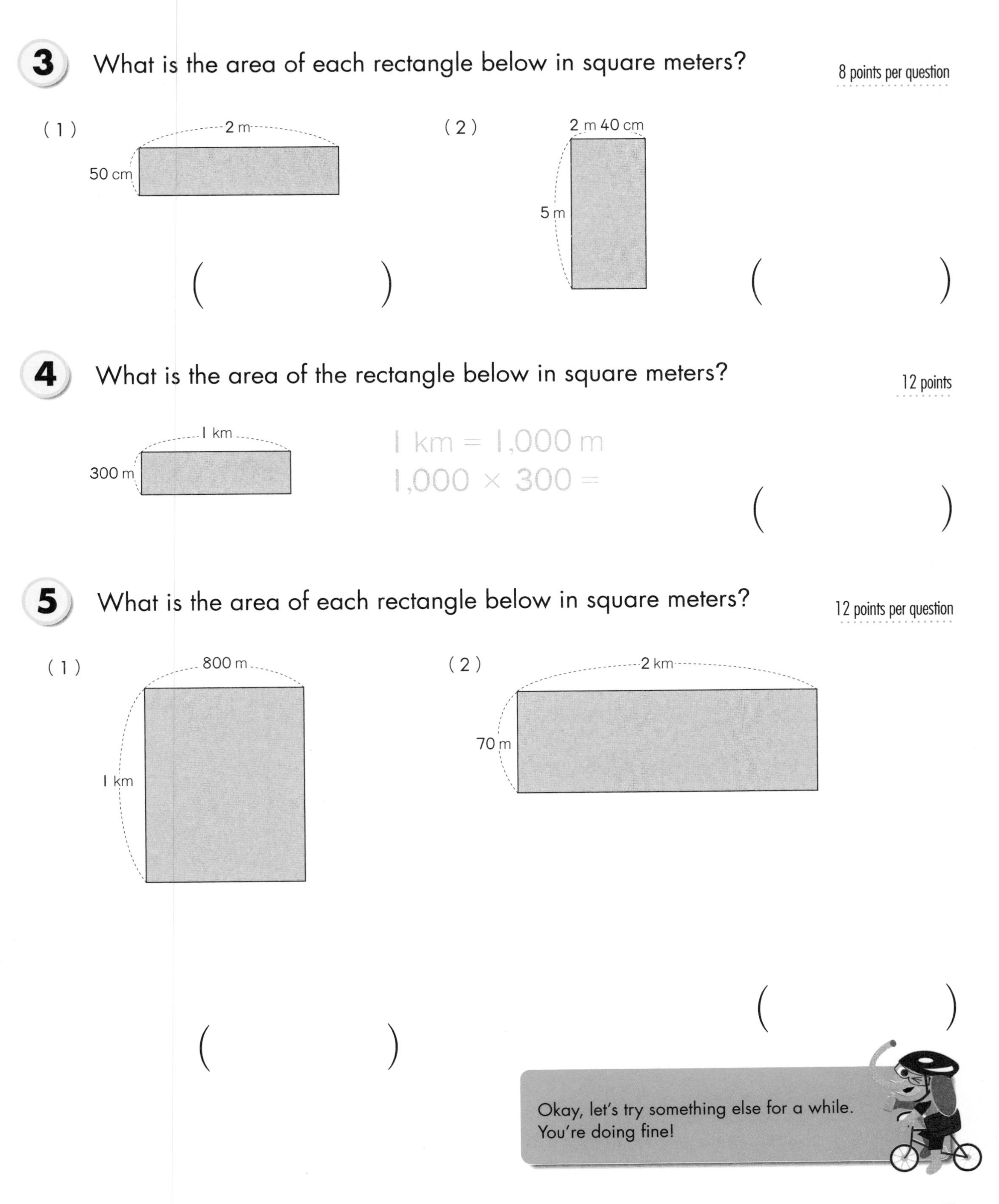

3 What is the area of each rectangle below in square meters?

8 points per question

(1) 2 m × 50 cm

( )

(2) 2 m 40 cm × 5 m

( )

4 What is the area of the rectangle below in square meters?

12 points

1 km × 300 m

1 km = 1,000 m
1,000 × 300 =

( )

5 What is the area of each rectangle below in square meters?

12 points per question

(1) 800 m × 1 km

( )

(2) 2 km × 70 m

( )

Okay, let's try something else for a while.
You're doing fine!

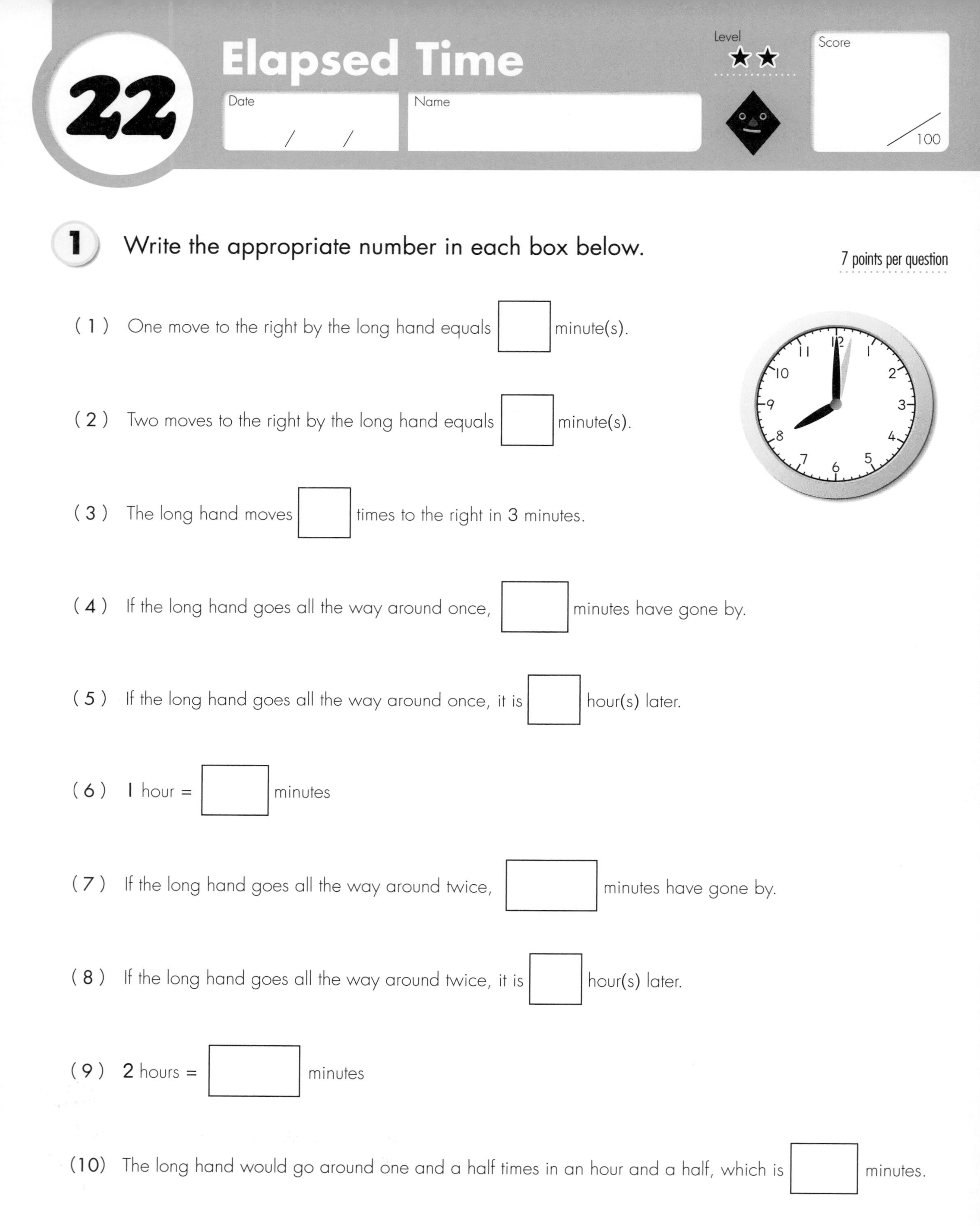

22 Elapsed Time

Level ★★

Date / / Name

Score /100

1 Write the appropriate number in each box below.

7 points per question

(1) One move to the right by the long hand equals ☐ minute(s).

(2) Two moves to the right by the long hand equals ☐ minute(s).

(3) The long hand moves ☐ times to the right in 3 minutes.

(4) If the long hand goes all the way around once, ☐ minutes have gone by.

(5) If the long hand goes all the way around once, it is ☐ hour(s) later.

(6) 1 hour = ☐ minutes

(7) If the long hand goes all the way around twice, ☐ minutes have gone by.

(8) If the long hand goes all the way around twice, it is ☐ hour(s) later.

(9) 2 hours = ☐ minutes

(10) The long hand would go around one and a half times in an hour and a half, which is ☐ minutes.

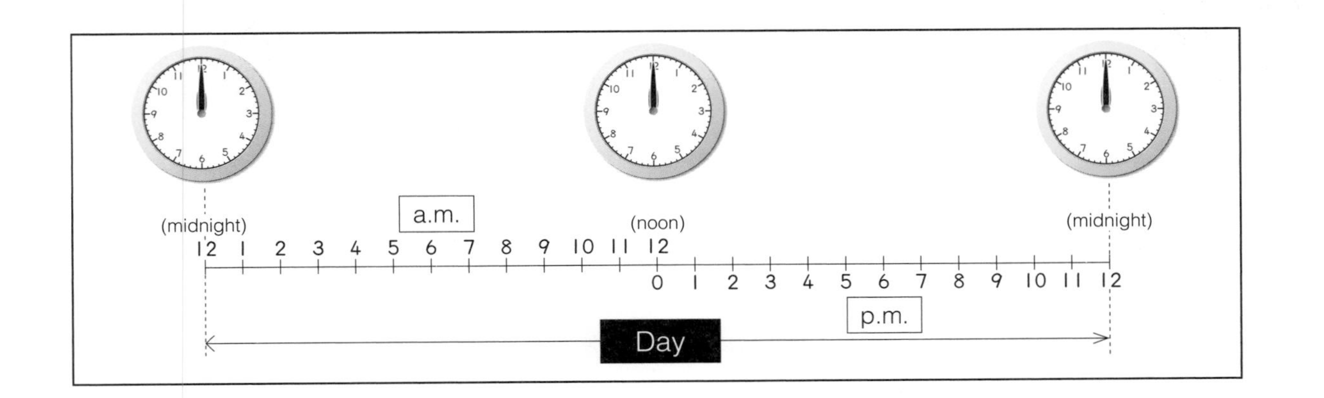

2 Use the figure above in order to answer the questions below.

5 points per question

(1) We describe the time from midnight to noon as [ ].

(2) We describe the time from noon to midnight as [ ].

(3) The amount of time that passes from midnight to noon is [ ] hours.

(4) The amount of time that passes from noon to midnight is [ ] hours.

(5) A day has [ ] hours in the morning and [ ] hours in the afternoon and evening.

(6) One day = [ ] hours

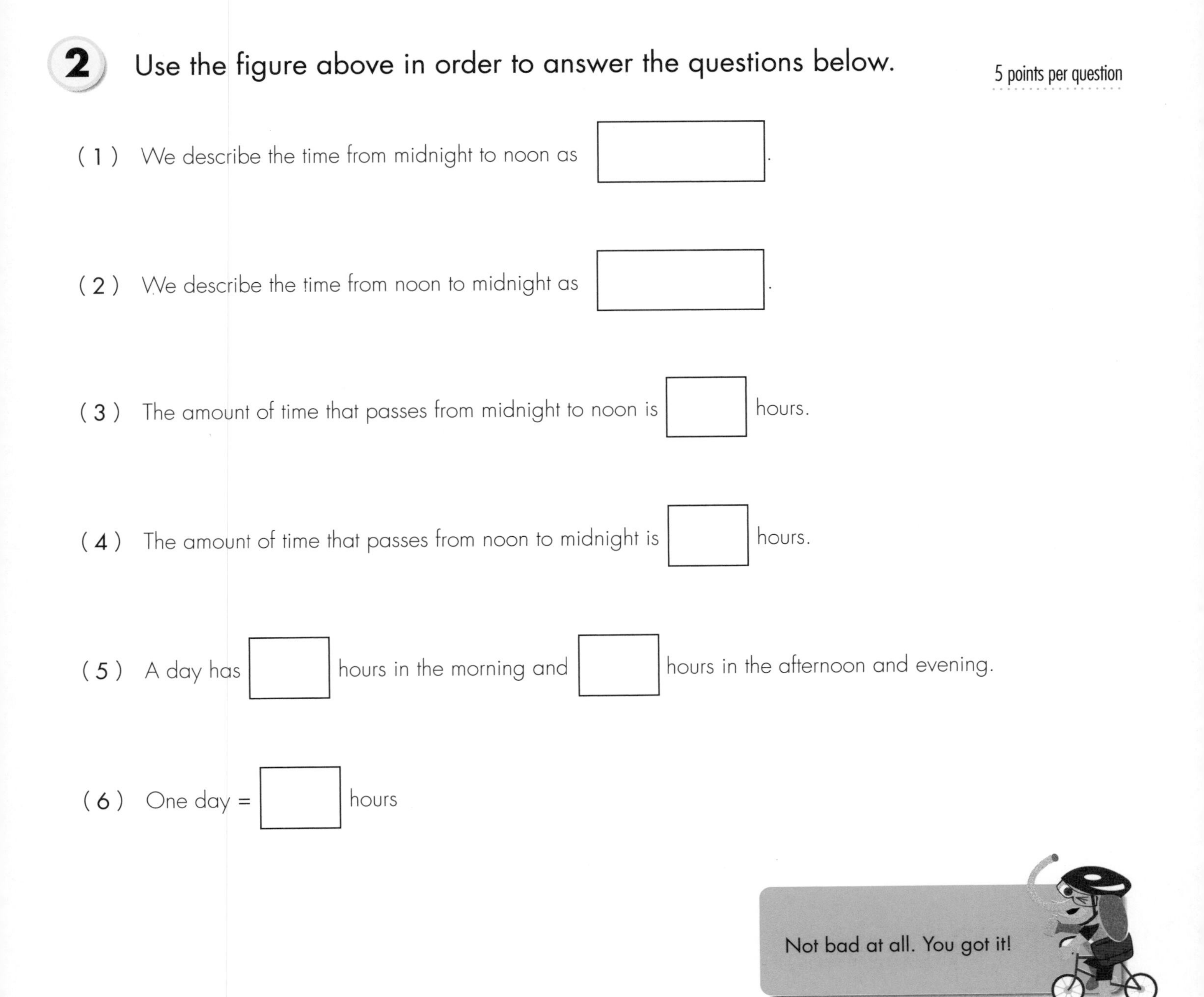

23 Elapsed Time

Level ★★

Date / /　Name

Score /100

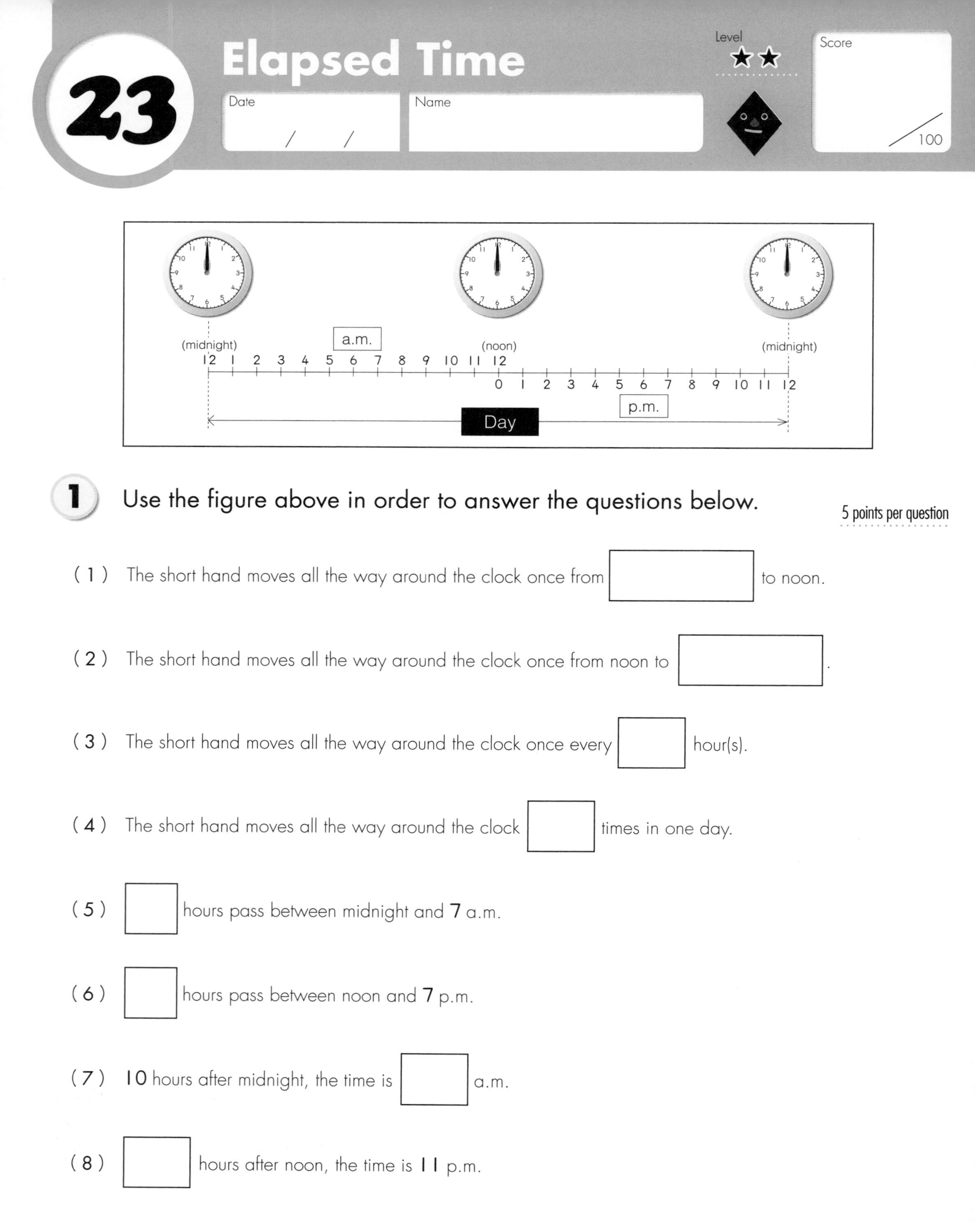

1 Use the figure above in order to answer the questions below.

5 points per question

(1) The short hand moves all the way around the clock once from ☐ to noon.

(2) The short hand moves all the way around the clock once from noon to ☐.

(3) The short hand moves all the way around the clock once every ☐ hour(s).

(4) The short hand moves all the way around the clock ☐ times in one day.

(5) ☐ hours pass between midnight and 7 a.m.

(6) ☐ hours pass between noon and 7 p.m.

(7) 10 hours after midnight, the time is ☐ a.m.

(8) ☐ hours after noon, the time is 11 p.m.

2 Write the time below each clock. Remember to include if it is a.m. or p.m.

6 points per question

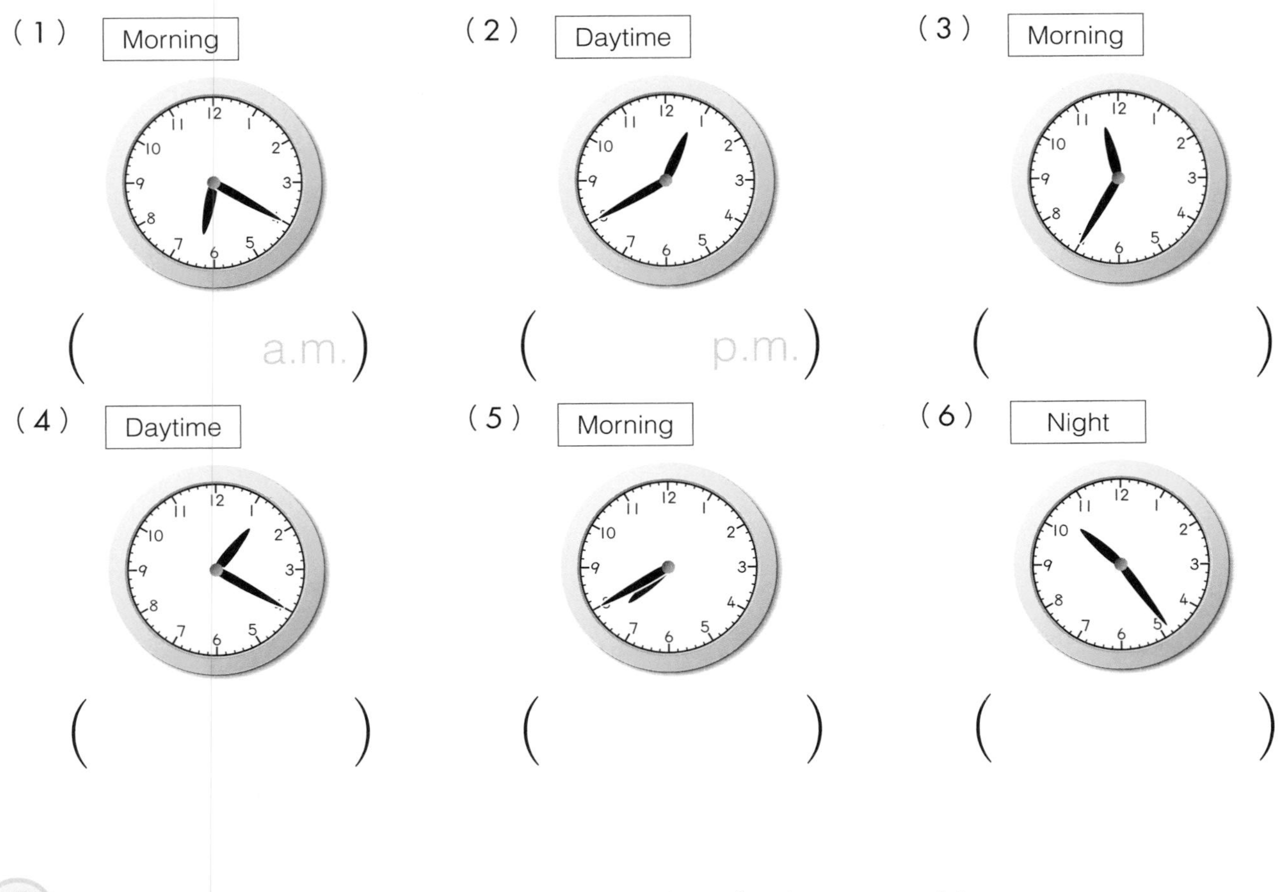

3 Answer the questions below using the clock pictured here.

6 points per question

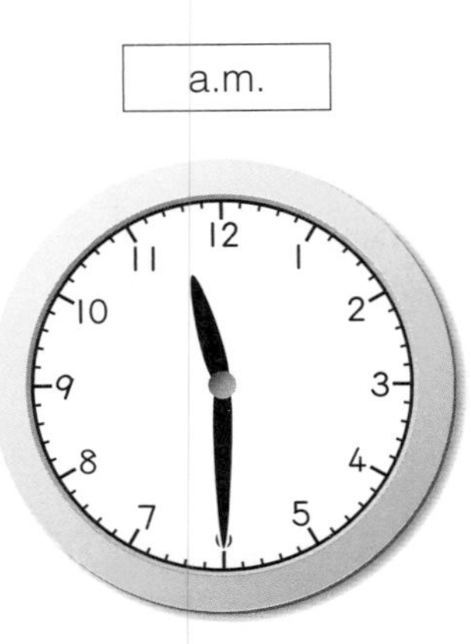

(1) Half an hour ago ( )

(2) Half an hour from now ( )

(3) An hour ago ( )

(4) An hour from now ( )

Does it feel like you've been working all day? Take a break!

24 Elapsed Time

Level ★★★

Date / /

Name

Score /100

1 How many minutes have passed from the time on the left to the time on the right?

5 points per question

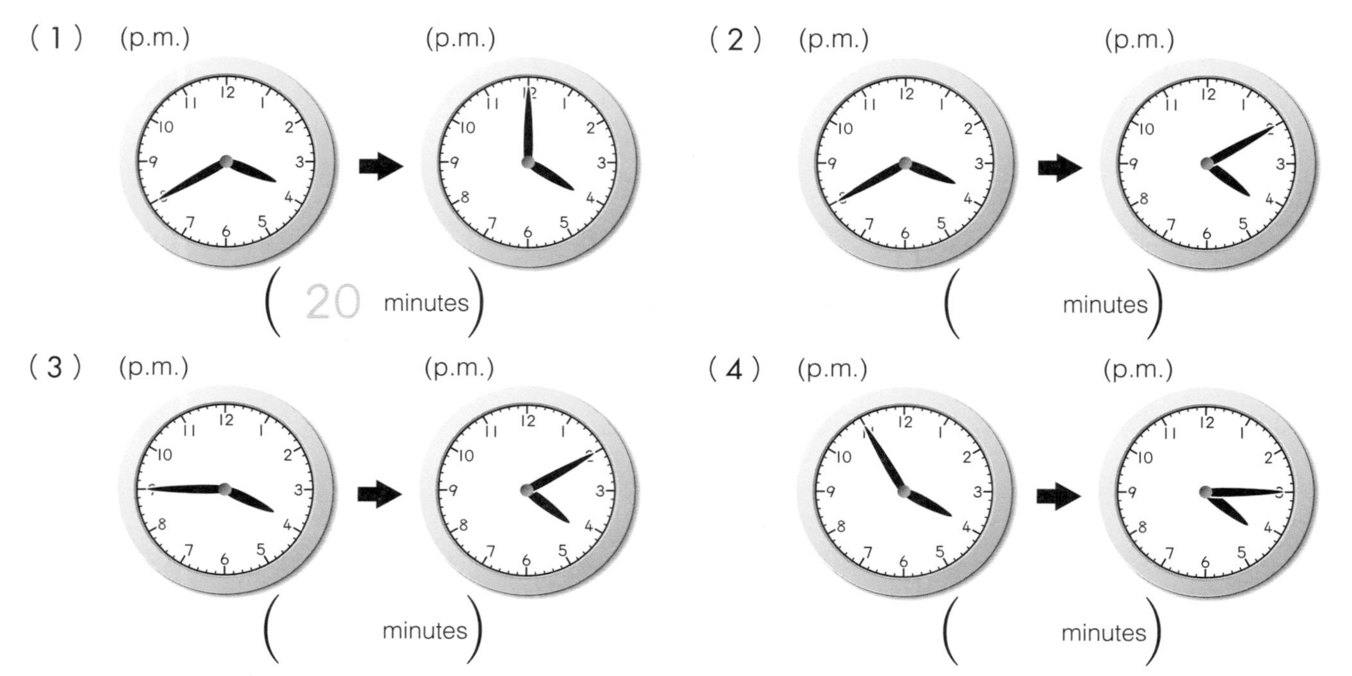

(1) (p.m.) → (p.m.) (20 minutes)

(2) (p.m.) → (p.m.) (minutes)

(3) (p.m.) → (p.m.) (minutes)

(4) (p.m.) → (p.m.) (minutes)

2 How much time has passed from the time on the left to the time on the right?

6 points per question

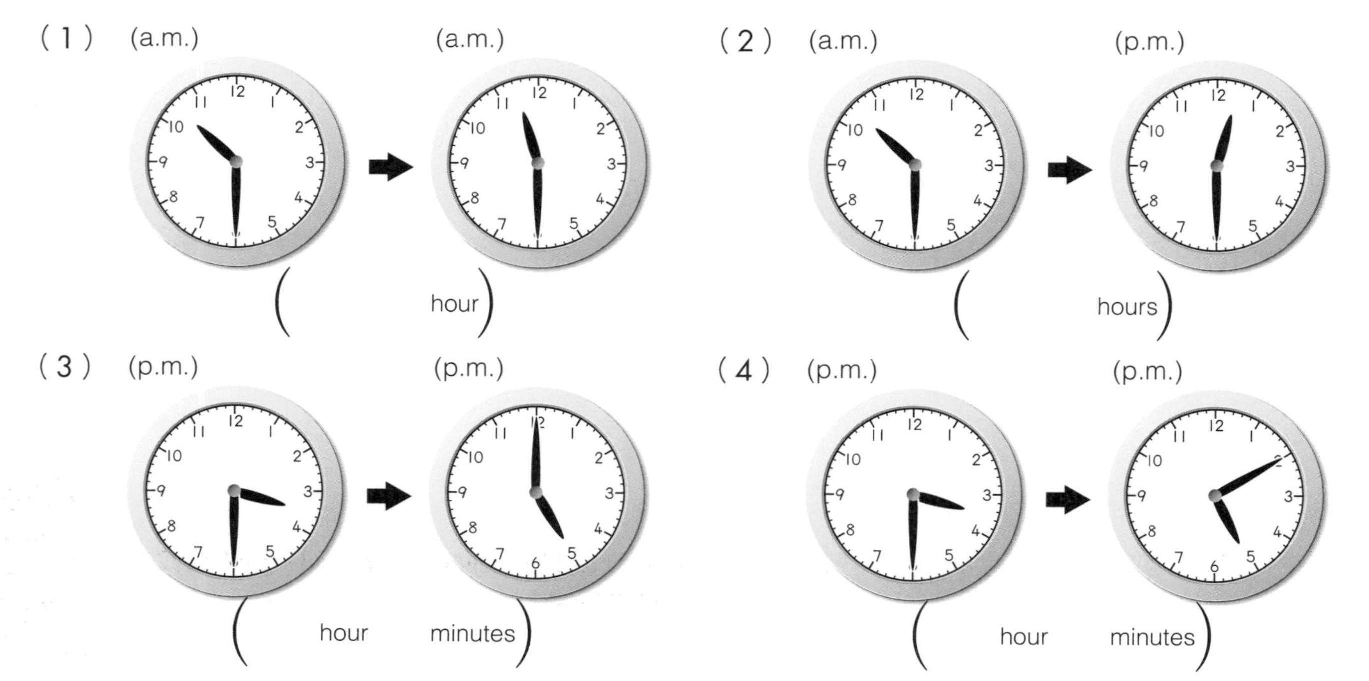

(1) (a.m.) → (a.m.) (hour)

(2) (a.m.) → (p.m.) (hours)

(3) (p.m.) → (p.m.) (hour minutes)

(4) (p.m.) → (p.m.) (hour minutes)

3 Read each clock below. Then add the amount of time in brackets to find the final time.

7 points per question

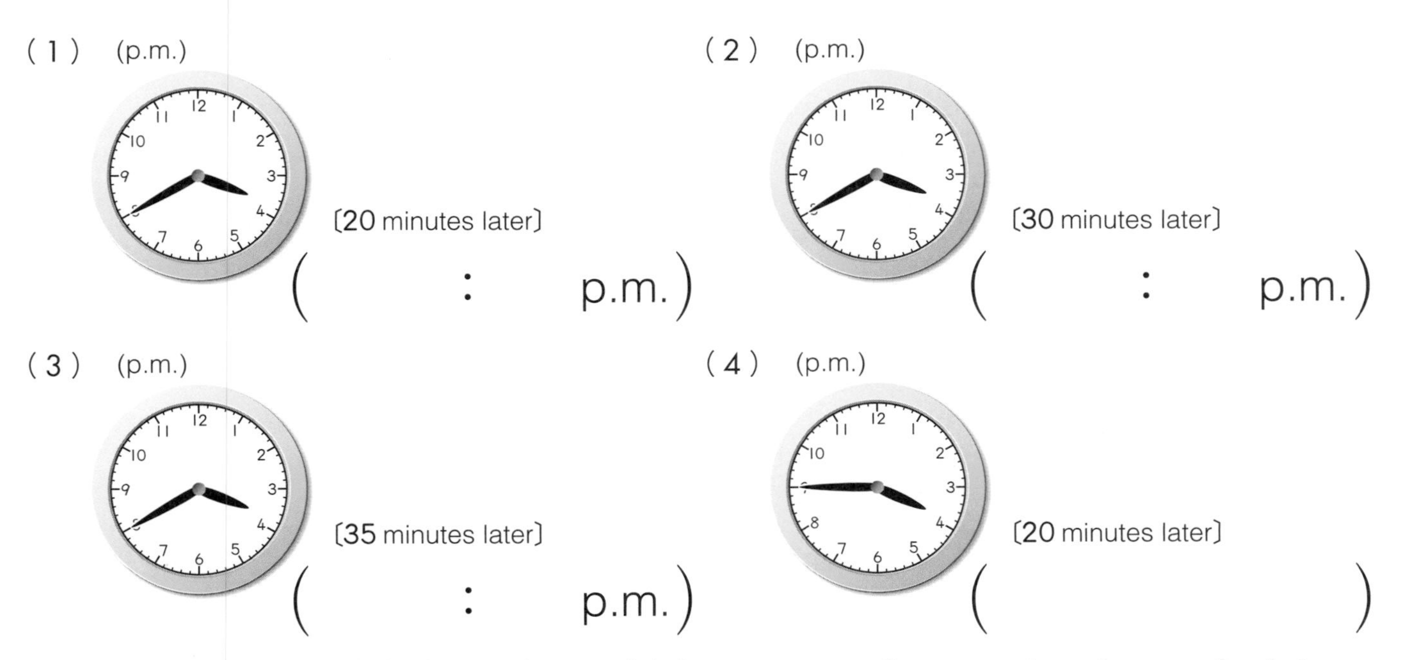

4 Read each clock below. Then add the amount of time in brackets to find the final time.

7 points per question

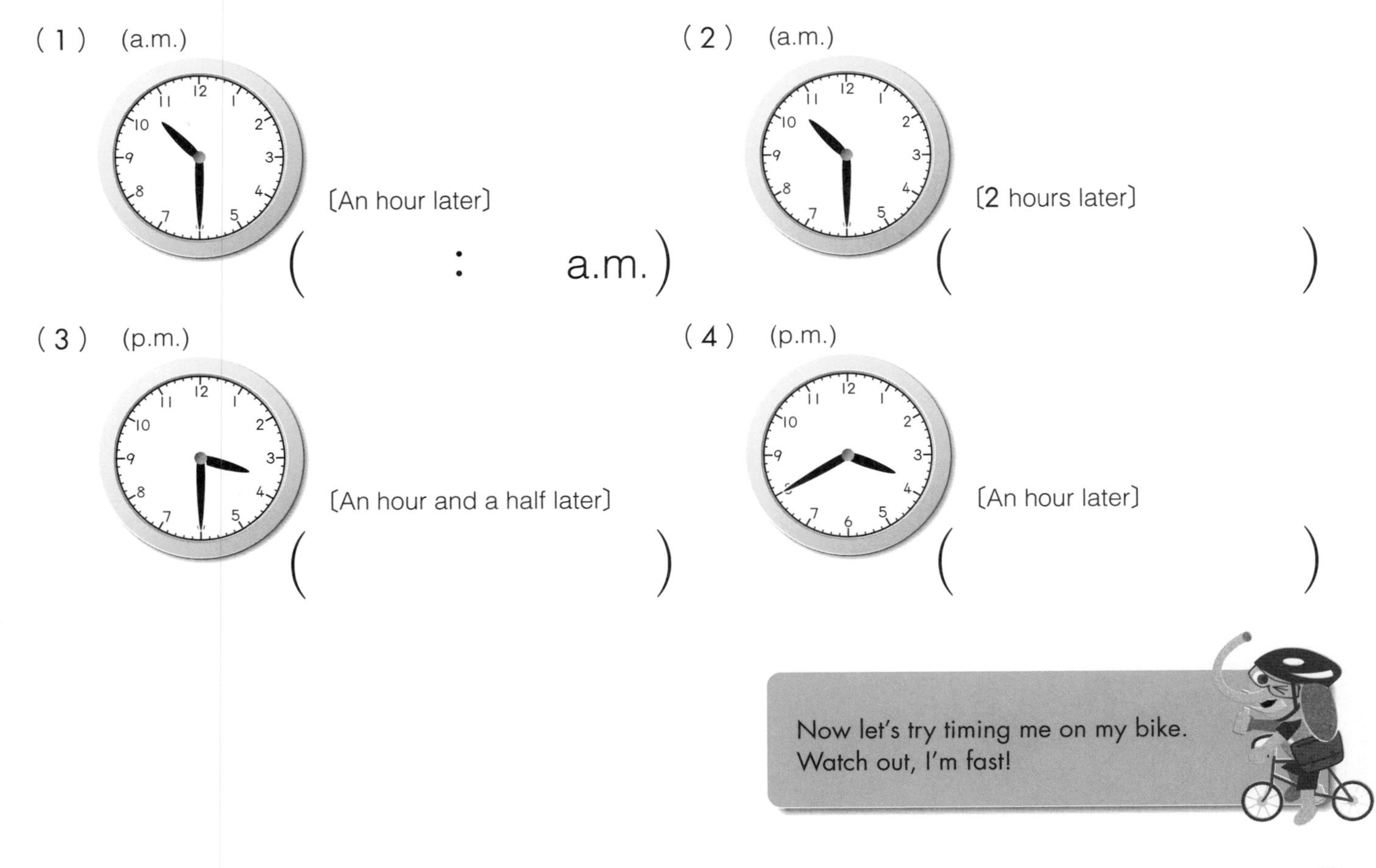

Now let's try timing me on my bike.
Watch out, I'm fast!

1 The track coach is timing us today in our events. The stopwatch he's using has a long hand that moves all the way around the face of the watch once every 60 seconds. How many seconds have gone by on each stopwatch below?

4 points per question

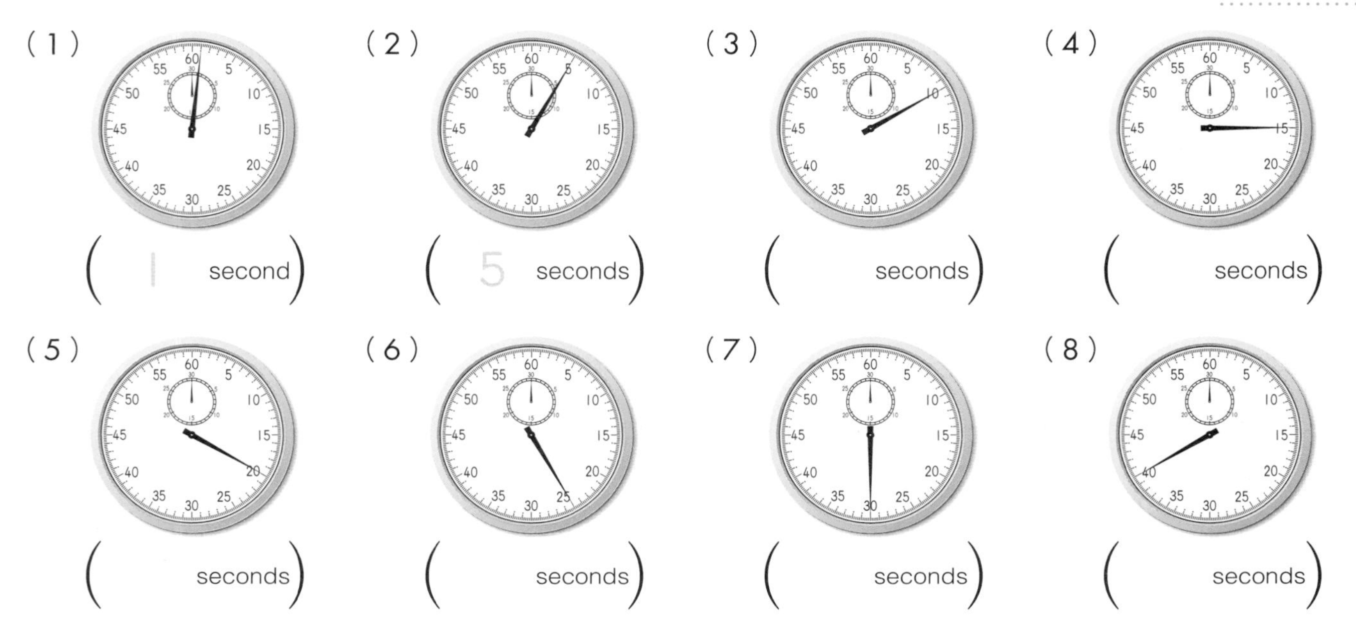

(1) (1 second)
(2) (5 seconds)
(3) (seconds)
(4) (seconds)
(5) (seconds)
(6) (seconds)
(7) (seconds)
(8) (seconds)

2 The long hand of his stopwatch goes all the way around once every 60 seconds, and the shorthand moves once every minute. How much time has gone by on each stopwatch below?

5 points per question

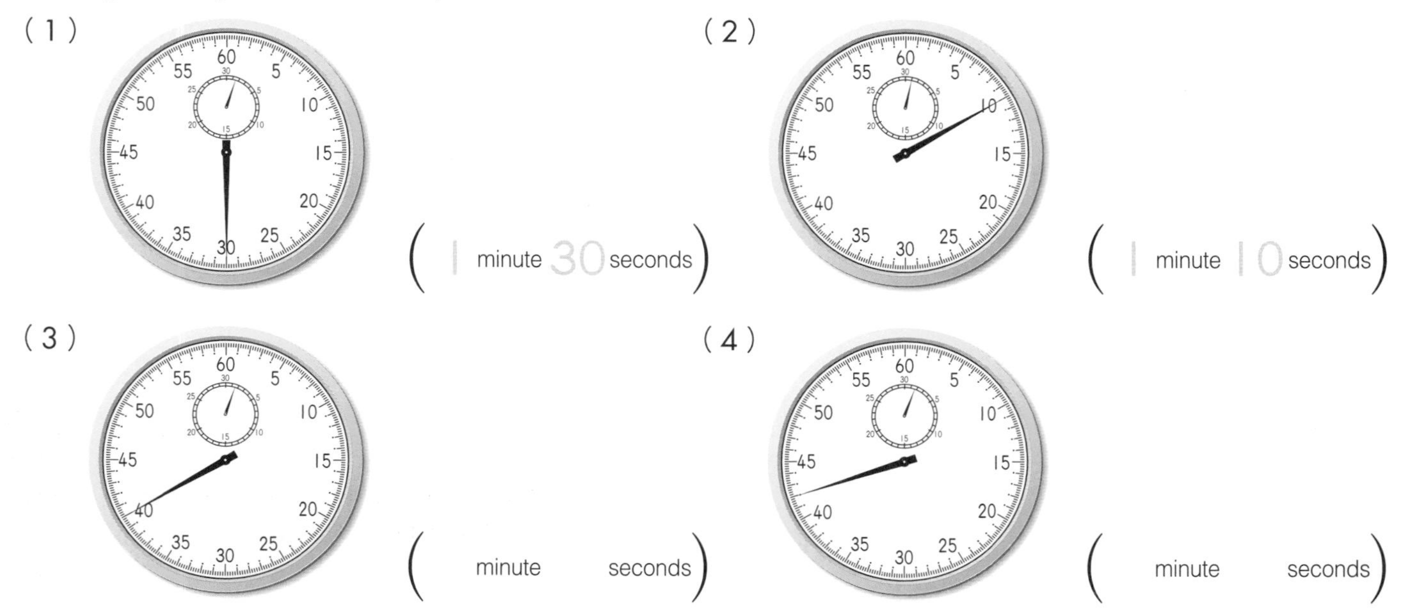

(1) (1 minute 30 seconds)
(2) (1 minute 10 seconds)
(3) (minute seconds)
(4) (minute seconds)

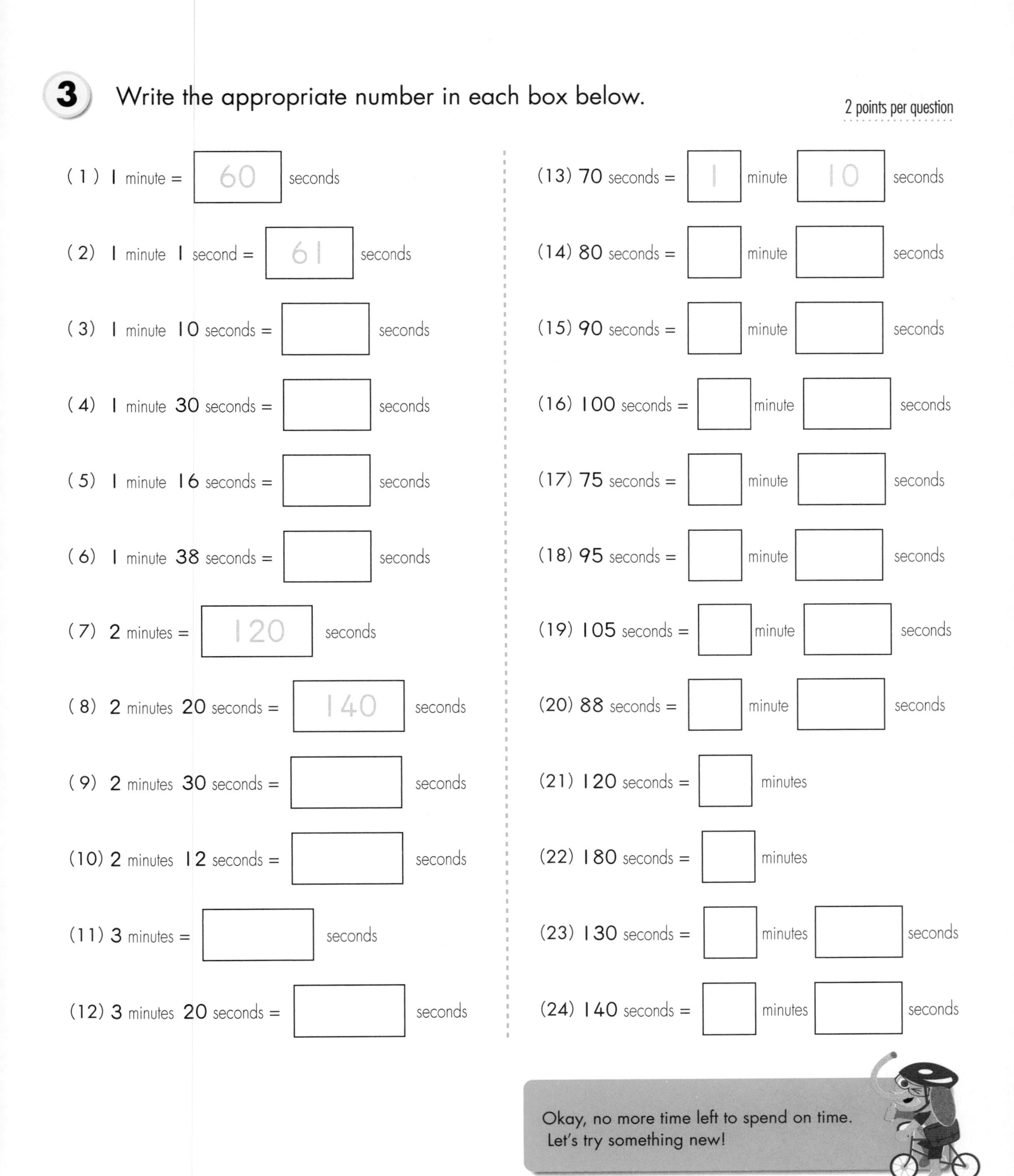

3 Write the appropriate number in each box below.

2 points per question

(1) 1 minute = [60] seconds

(2) 1 minute 1 second = [61] seconds

(3) 1 minute 10 seconds = [ ] seconds

(4) 1 minute 30 seconds = [ ] seconds

(5) 1 minute 16 seconds = [ ] seconds

(6) 1 minute 38 seconds = [ ] seconds

(7) 2 minutes = [120] seconds

(8) 2 minutes 20 seconds = [140] seconds

(9) 2 minutes 30 seconds = [ ] seconds

(10) 2 minutes 12 seconds = [ ] seconds

(11) 3 minutes = [ ] seconds

(12) 3 minutes 20 seconds = [ ] seconds

(13) 70 seconds = [1] minute [10] seconds

(14) 80 seconds = [ ] minute [ ] seconds

(15) 90 seconds = [ ] minute [ ] seconds

(16) 100 seconds = [ ] minute [ ] seconds

(17) 75 seconds = [ ] minute [ ] seconds

(18) 95 seconds = [ ] minute [ ] seconds

(19) 105 seconds = [ ] minute [ ] seconds

(20) 88 seconds = [ ] minute [ ] seconds

(21) 120 seconds = [ ] minutes

(22) 180 seconds = [ ] minutes

(23) 130 seconds = [ ] minutes [ ] seconds

(24) 140 seconds = [ ] minutes [ ] seconds

Okay, no more time left to spend on time. Let's try something new!

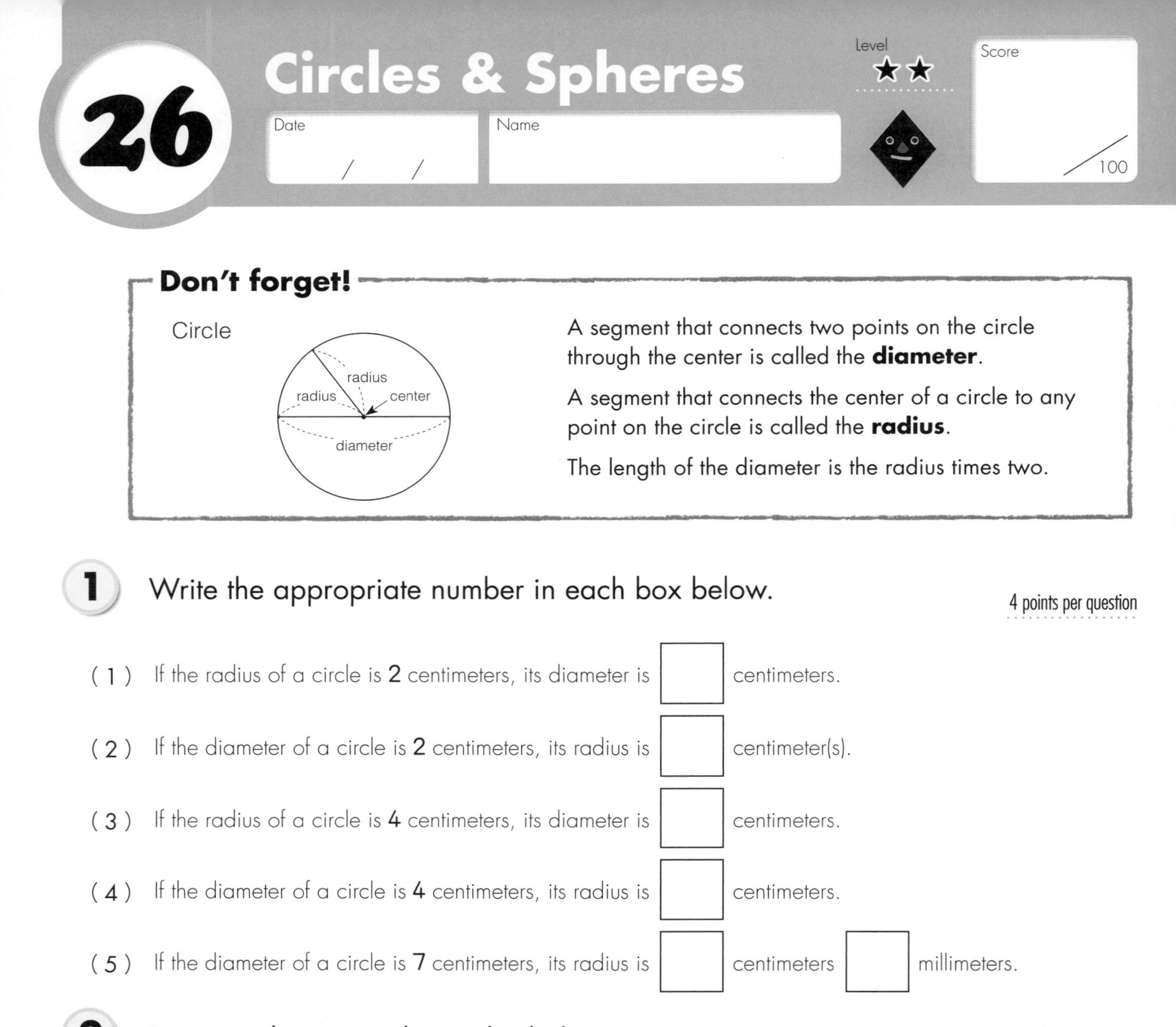

26 Circles & Spheres

Level ★★

Date / /

Name

Score /100

Don't forget!

Circle

A segment that connects two points on the circle through the center is called the **diameter**.

A segment that connects the center of a circle to any point on the circle is called the **radius**.

The length of the diameter is the radius times two.

1 Write the appropriate number in each box below.

4 points per question

(1) If the radius of a circle is **2** centimeters, its diameter is ☐ centimeters.

(2) If the diameter of a circle is **2** centimeters, its radius is ☐ centimeter(s).

(3) If the radius of a circle is **4** centimeters, its diameter is ☐ centimeters.

(4) If the diameter of a circle is **4** centimeters, its radius is ☐ centimeters.

(5) If the diameter of a circle is **7** centimeters, its radius is ☐ centimeters ☐ millimeters.

2 Draw each item on the circles below.

10 points per question

(1) Diameter

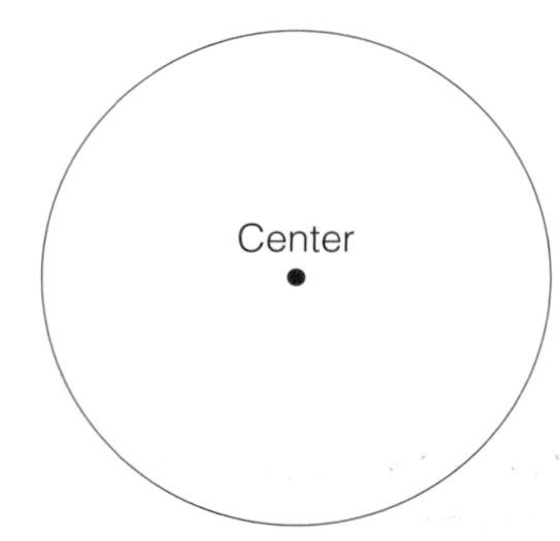

(2) Radius

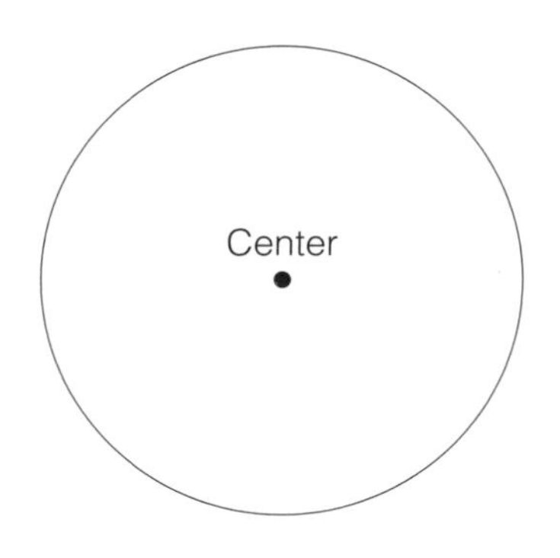

Use a compass to draw each circle below.

10 points per question

(1) This circle has a radius of **3** centimeters.

(2) This circle has a diameter of **6** centimeters.

(3) This circle has a radius of **2** centimeters.

(4) This circle has a diameter of **4** centimeters.

(5) This circle has a diameter of **3** centimeters.

(6) This circle has a diameter of **5** centimeters.

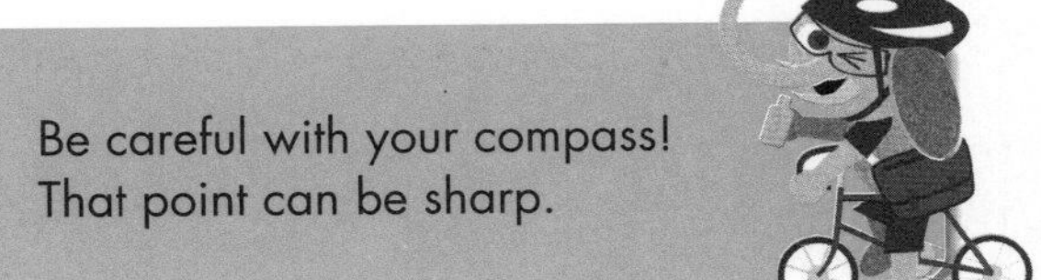

27 Circles & Spheres

1 Two circles of the same size are inside a larger circle that has a diameter of 8 inches.

5 points per question

(1) What is the radius of the big circle?

( )

(2) What is the diameter of each small circle?

( )

(3) What is the radius of each small circle?

( )

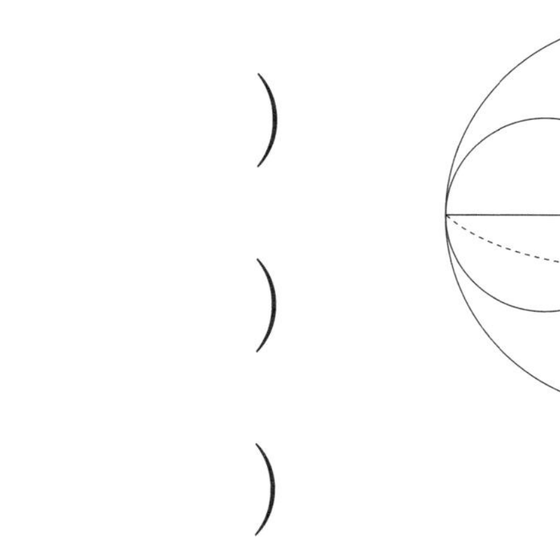

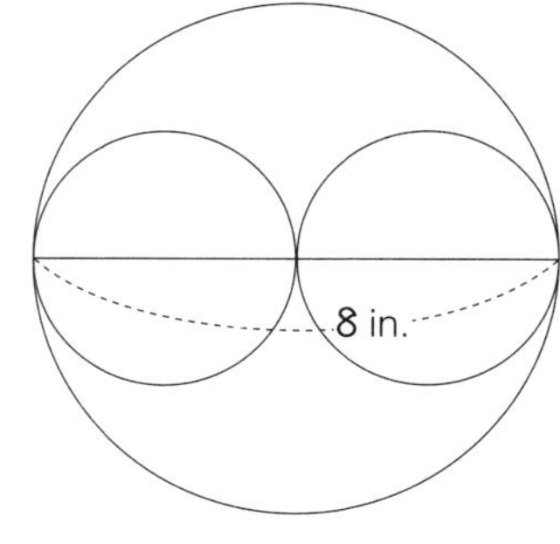

2 Two circles of the same size are inside a larger circle that has a radius of 5 inches.

5 points per question

(1) What is the diameter of the big circle?

( )

(2) What is the diameter of each small circle?

( )

(3) What is the radius of each small circle?

( )

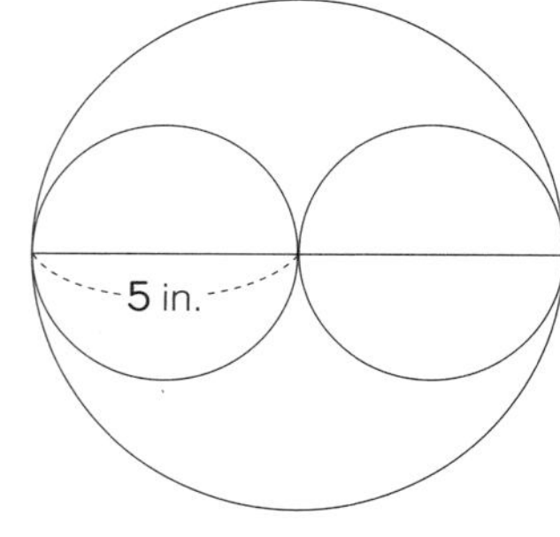

3 The radius of the circle on the right is 3 inches.

5 points per question

(1) What is the diameter of the circle?

( )

(2) How long is each side of the square?

( )

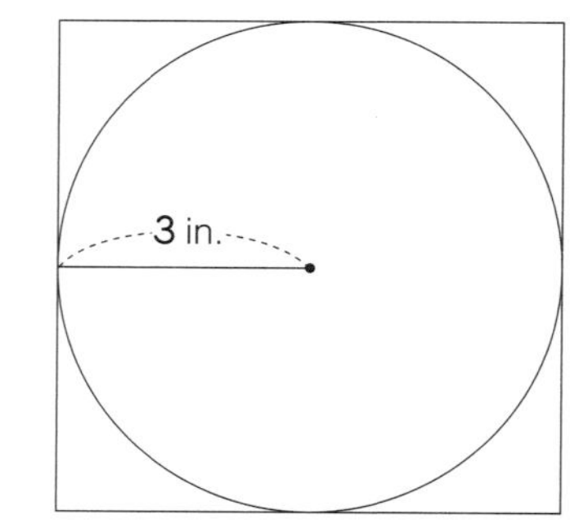

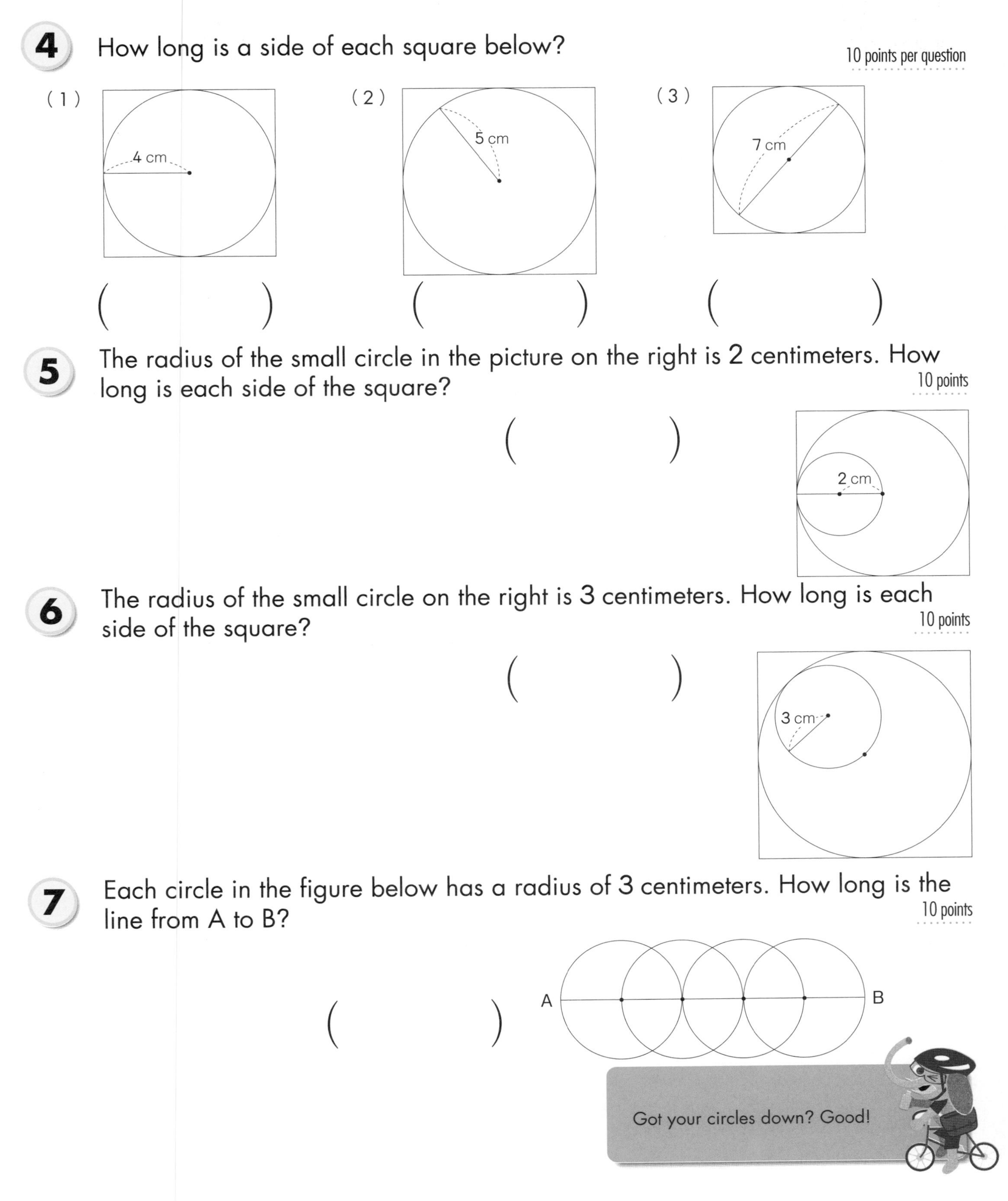

4 How long is a side of each square below?

10 points per question

(1) 4 cm ( )

(2) 5 cm ( )

(3) 7 cm ( )

5 The radius of the small circle in the picture on the right is 2 centimeters. How long is each side of the square?

10 points

( )

2 cm

6 The radius of the small circle on the right is 3 centimeters. How long is each side of the square?

10 points

( )

3 cm

7 Each circle in the figure below has a radius of 3 centimeters. How long is the line from A to B?

10 points

( )

A B

Got your circles down? Good!

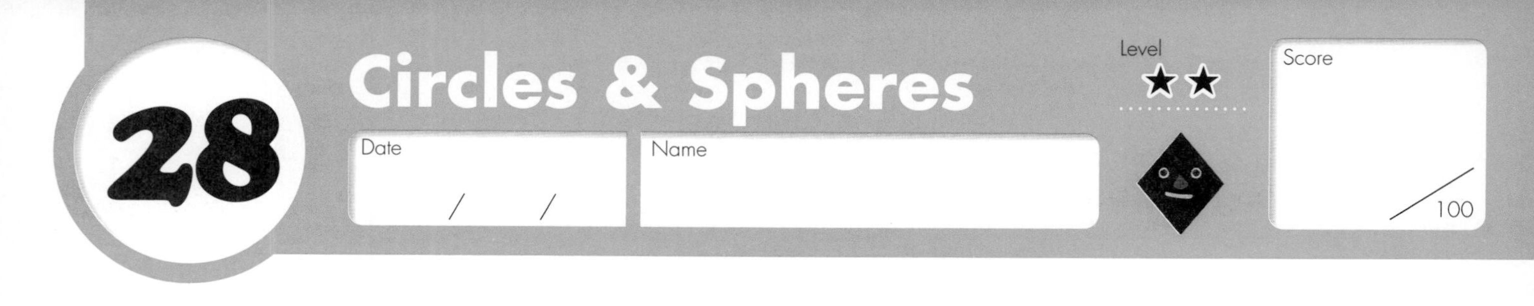

28 Circles & Spheres

1 If you cut each baseball below as pictured, what shape will the cross section have? Connect each ball to the correct shape.

15 points for completion

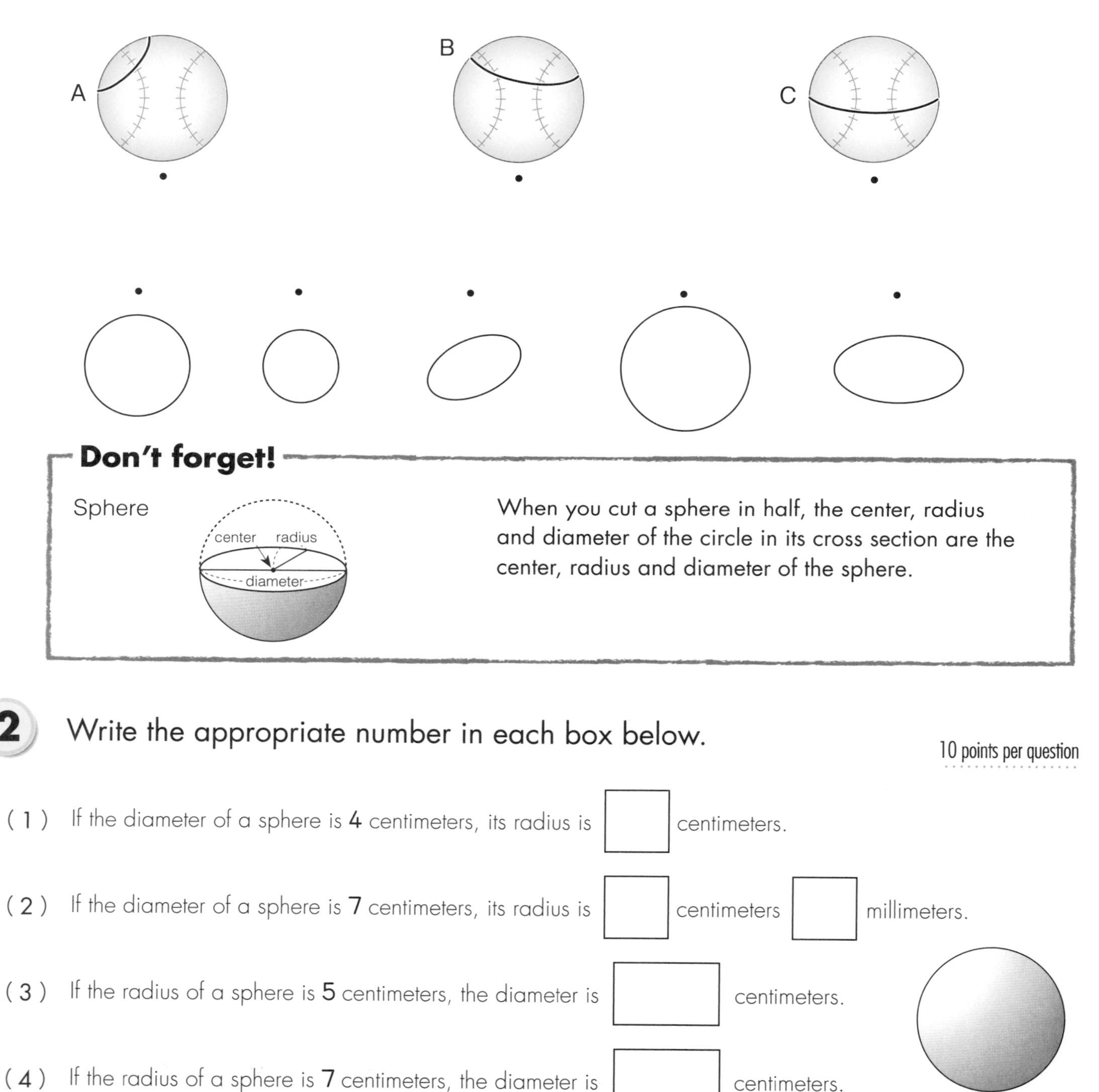

Don't forget!

When you cut a sphere in half, the center, radius and diameter of the circle in its cross section are the center, radius and diameter of the sphere.

2 Write the appropriate number in each box below.

10 points per question

(1) If the diameter of a sphere is 4 centimeters, its radius is ☐ centimeters.

(2) If the diameter of a sphere is 7 centimeters, its radius is ☐ centimeters ☐ millimeters.

(3) If the radius of a sphere is 5 centimeters, the diameter is ☐ centimeters.

(4) If the radius of a sphere is 7 centimeters, the diameter is ☐ centimeters.

Sphere

3 As pictured on the right, you have two blocks surrounding a ball.

5 points per question

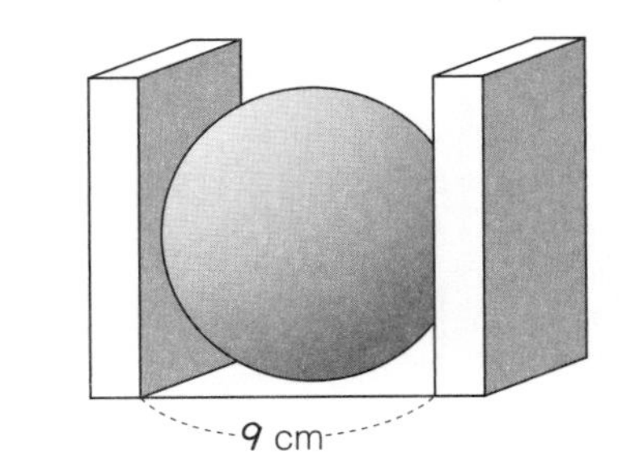

(1) What is the diameter of the ball?

(   )

(2) What is the radius of the ball?

(   )

4 As pictured on the right, you have a sphere that fits snugly inside a box.

5 points per question

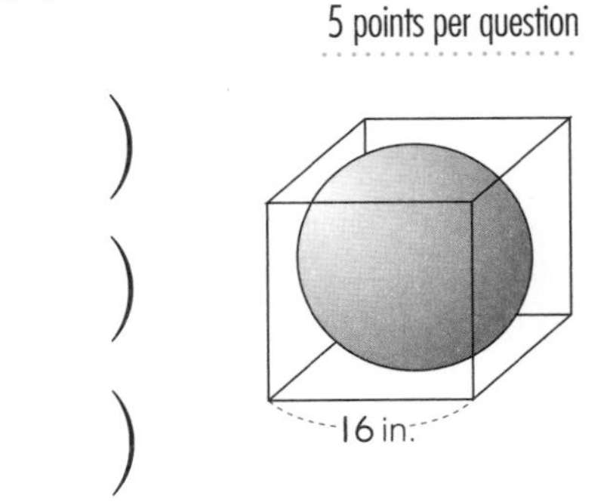

(1) How long is each side of the box? (   )

(2) How long is the diameter of the sphere? (   )

(3) How long is the radius of the sphere? (   )

5 As pictured on the right, you have three similar balls that fit snugly inside one box that is 24 inches wide.

5 points per question

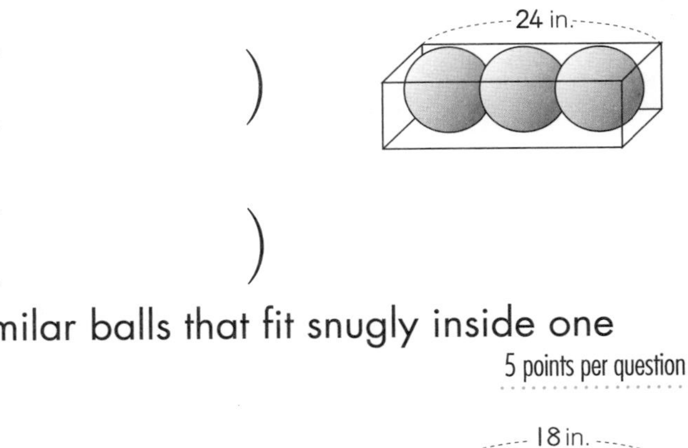

(1) How long is the diameter of each ball?

(   )

(2) How long is the radius of each ball?

(   )

6 As pictured on the right, you have six similar balls that fit snugly inside one box that is 18 inches wide.

5 points per question

(1) What is the diameter of each ball?

(   )

(2) What is the length of the box?

(   )

Spheres? No problem.
Let's see if you can handle triangles!

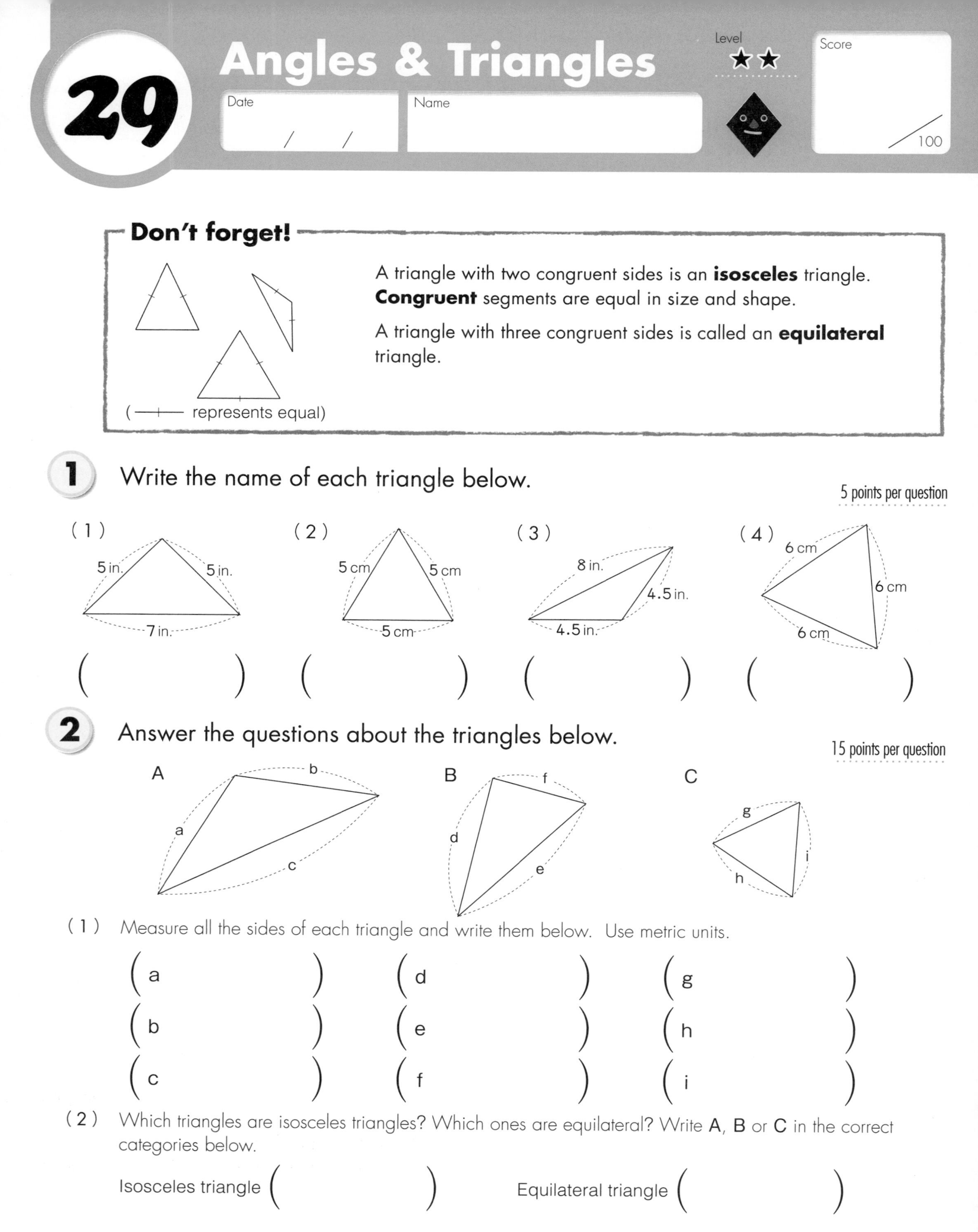

Don't forget!

A triangle with two congruent sides is an **isosceles** triangle. **Congruent** segments are equal in size and shape.

A triangle with three congruent sides is called an **equilateral** triangle.

(―+― represents equal)

1 Write the name of each triangle below.

5 points per question

(1) (　　　) (2) (　　　) (3) (　　　) (4) (　　　)

2 Answer the questions about the triangles below.

15 points per question

(1) Measure all the sides of each triangle and write them below. Use metric units.

(a　　　)	(d　　　)	(g　　　)
(b　　　)	(e　　　)	(h　　　)
(c　　　)	(f　　　)	(i　　　)

(2) Which triangles are isosceles triangles? Which ones are equilateral? Write A, B or C in the correct categories below.

Isosceles triangle (　　　) Equilateral triangle (　　　)

3 Sort the triangles below into equilateral and isosceles triangles by putting each letter next to the correct category. *10 points per question*

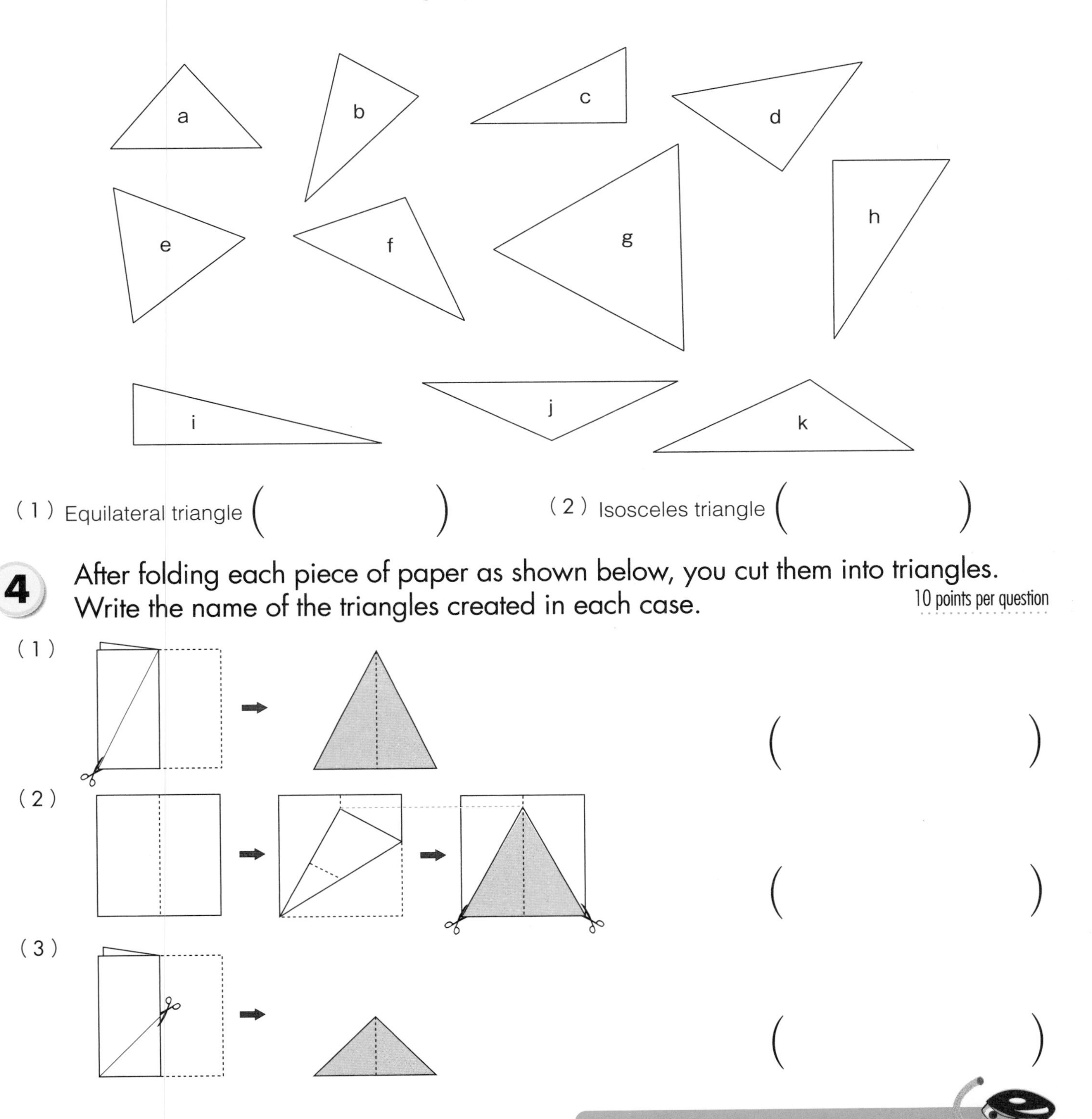

(1) Equilateral triangle (　　　　　)　　(2) Isosceles triangle (　　　　　)

4 After folding each piece of paper as shown below, you cut them into triangles. Write the name of the triangles created in each case. *10 points per question*

(1) (　　　　　)

(2) (　　　　　)

(3) (　　　　　)

Not so bad, right? Let's have some fun drawing our own triangles now!

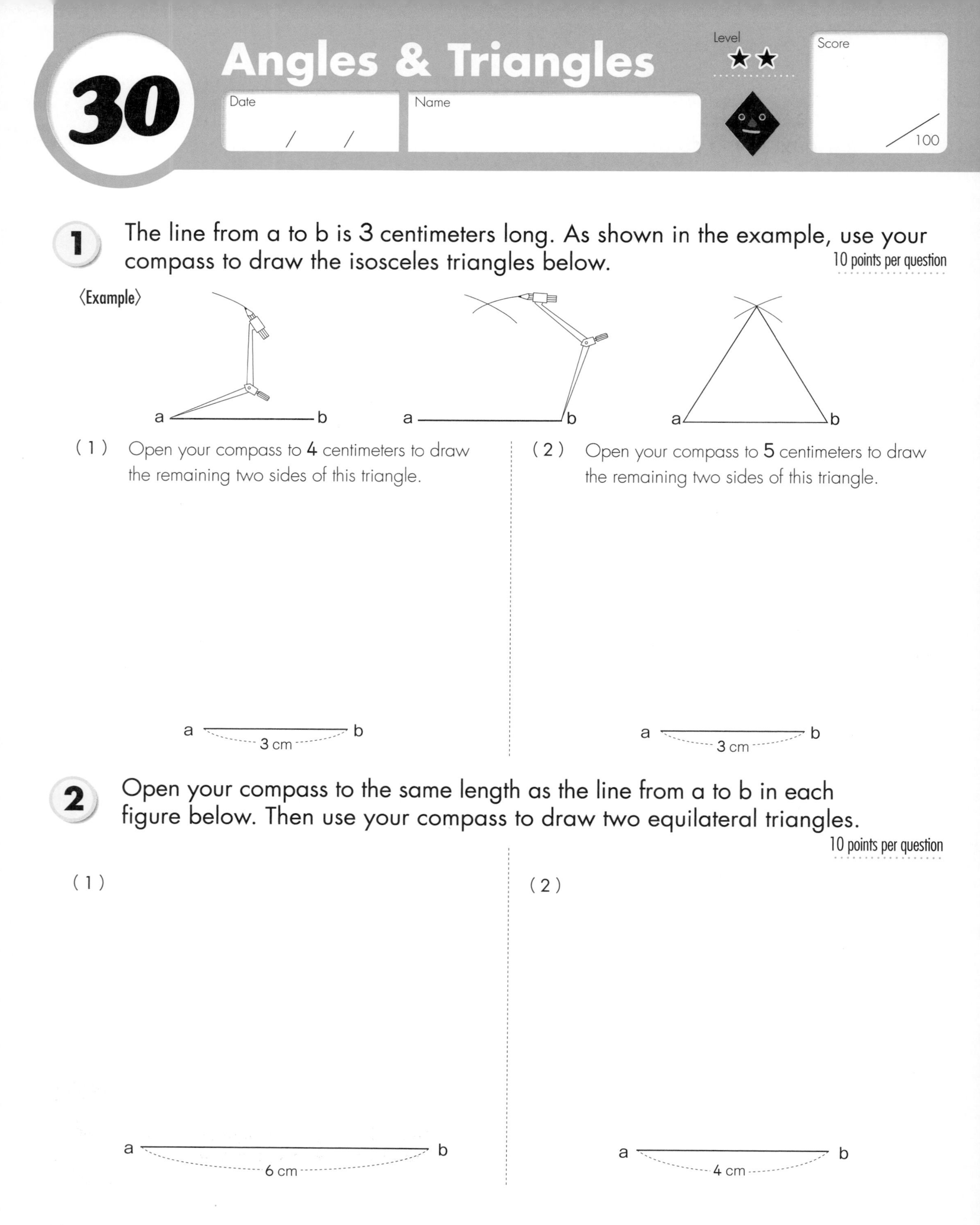

30 Angles & Triangles

Level ★★

Date / /

Name

Score /100

1 The line from a to b is 3 centimeters long. As shown in the example, use your compass to draw the isosceles triangles below.

10 points per question

〈Example〉

(1) Open your compass to 4 centimeters to draw the remaining two sides of this triangle.

(2) Open your compass to 5 centimeters to draw the remaining two sides of this triangle.

2 Open your compass to the same length as the line from a to b in each figure below. Then use your compass to draw two equilateral triangles.

10 points per question

(1)

(2)

3 Use your ruler and compass to draw the triangles below.

10 points per question

(1) This triangle has three sides that are 4 centimeters, 5 centimeters, and 4 centimeters long.

(2) This triangle has three sides that are 5 centimeters, 2 centimeters, and 5 centimeters long.

(3) This triangle has three sides that are 5 centimeters, 5 centimeters, and 5 centimeters long.

(4) This triangle has three sides that are all 3 centimeters 5 millimeters long.

4 The circle shown in the figure below has a radius of 3 centimeters.

5 points per question

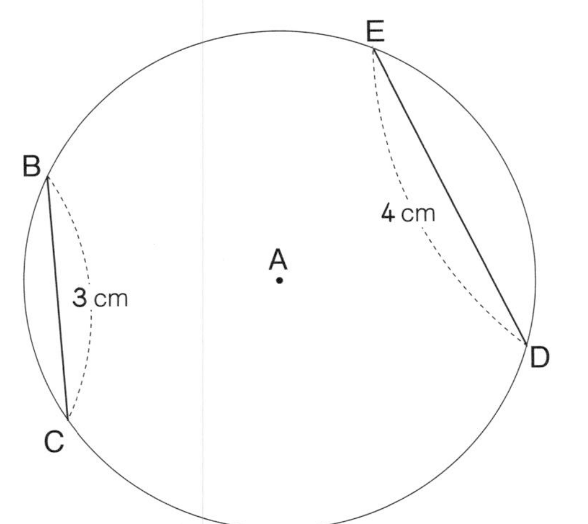

(1) Draw a triangle that has the vertices A, B and C.

(2) What kind of triangle is ABC?

( )

(3) Draw a triangle that has the vertices A, D and E.

(4) What kind of triangle is ADE?

( )

You're doing great. Keep it up!

Don't forget!

An **angle** is the geometric figure formed by two distinct rays that have one common endpoint. This common endpoint is called the **vertex** and the rays are called the **sides** of the angle.

The measure of an angle is the size of the space between those two distinct rays that have one common end point.

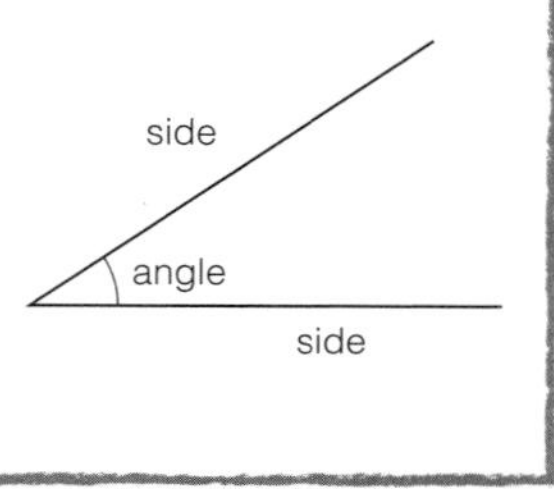

1 Use your triangular ruler in order to answer the questions below. 5 points per question

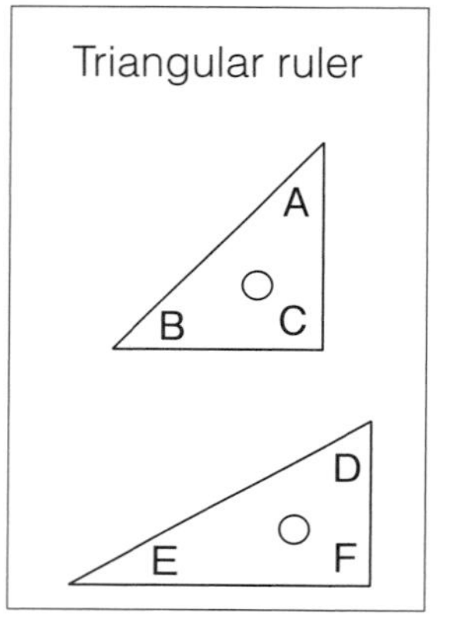

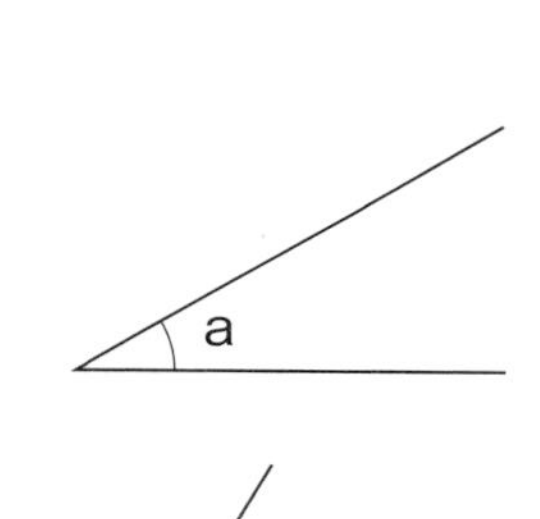

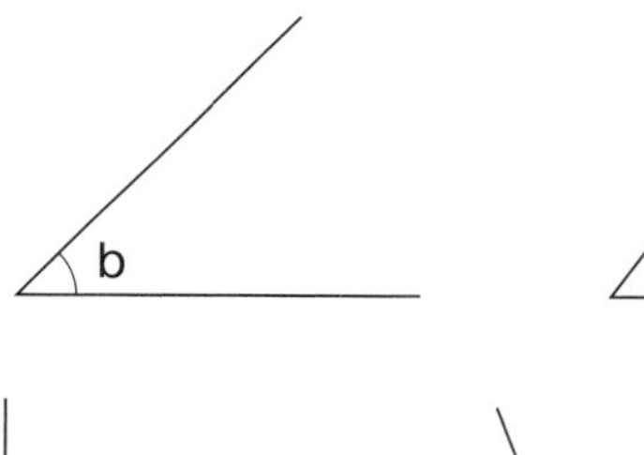

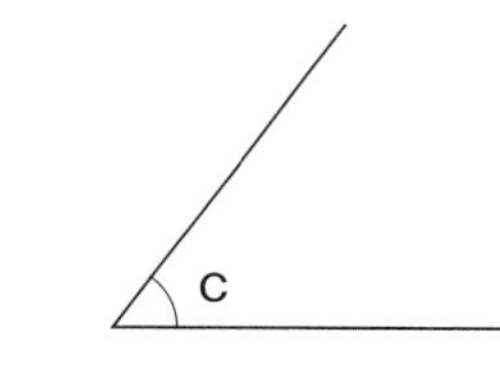

(1) Which angle is the same as angle A in the figure? ( )

(2) Which angle is the same as angle B in the figure? ( )

(3) Which angle is the same as angle C in the figure? ( )

(4) Which angle is the same as angle D in the figure? ( )

(5) Which angle is the same as angle E in the figure? ( )

(6) Which angle is the same as angle F in the figure? ( )

2 Rank the following angles from largest to smallest.

20 points per question

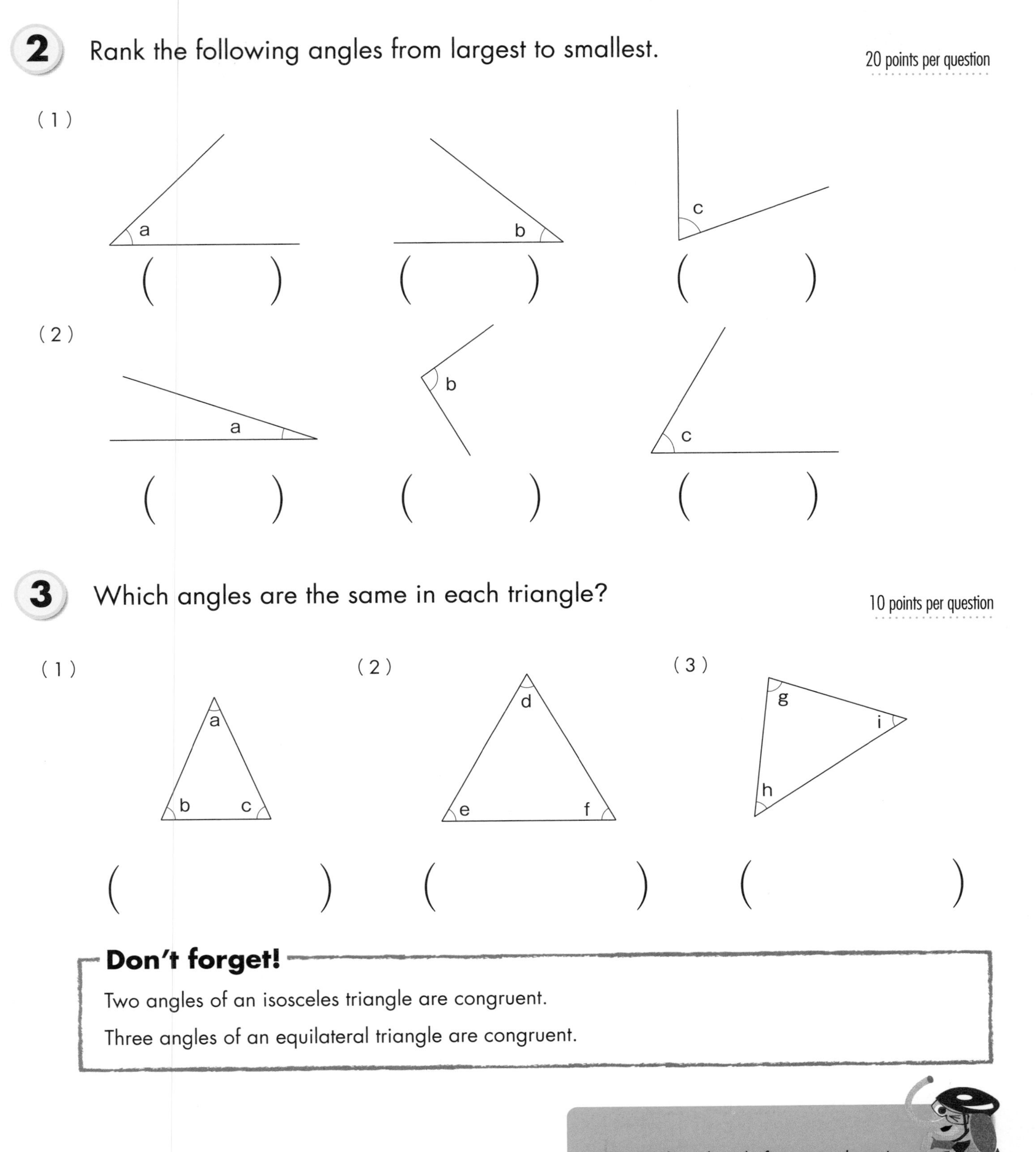

3 Which angles are the same in each triangle?

10 points per question

Don't forget!

Two angles of an isosceles triangle are congruent.

Three angles of an equilateral triangle are congruent.

Phew. Take a break if you need one!

32 Angles & Triangles

Level ★★

Date / /

Name

Score /100

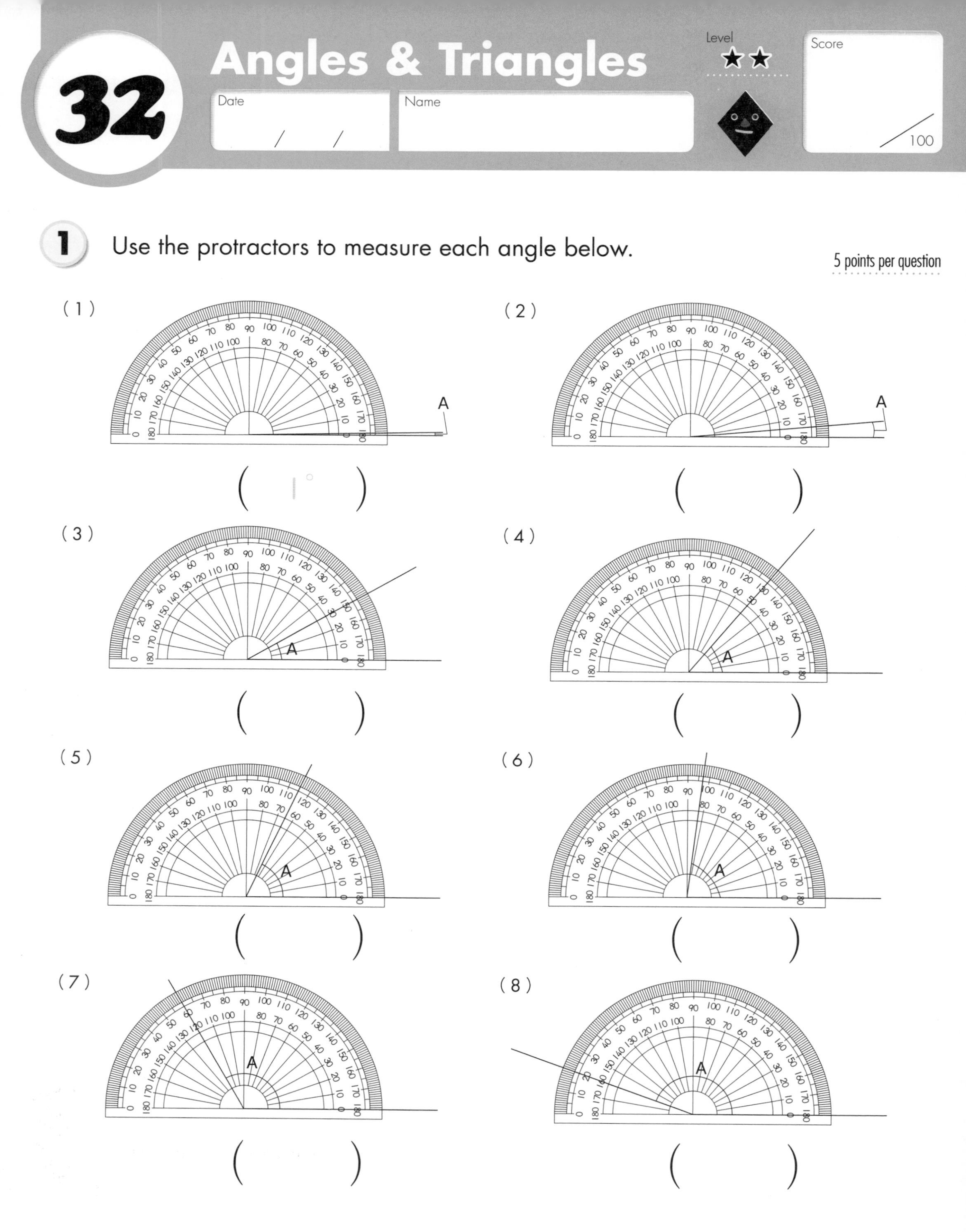

1 Use the protractors to measure each angle below.

5 points per question

(1) (1°)

(2) ( )

(3) ( )

(4) ( )

(5) ( )

(6) ( )

(7) ( )

(8) ( )

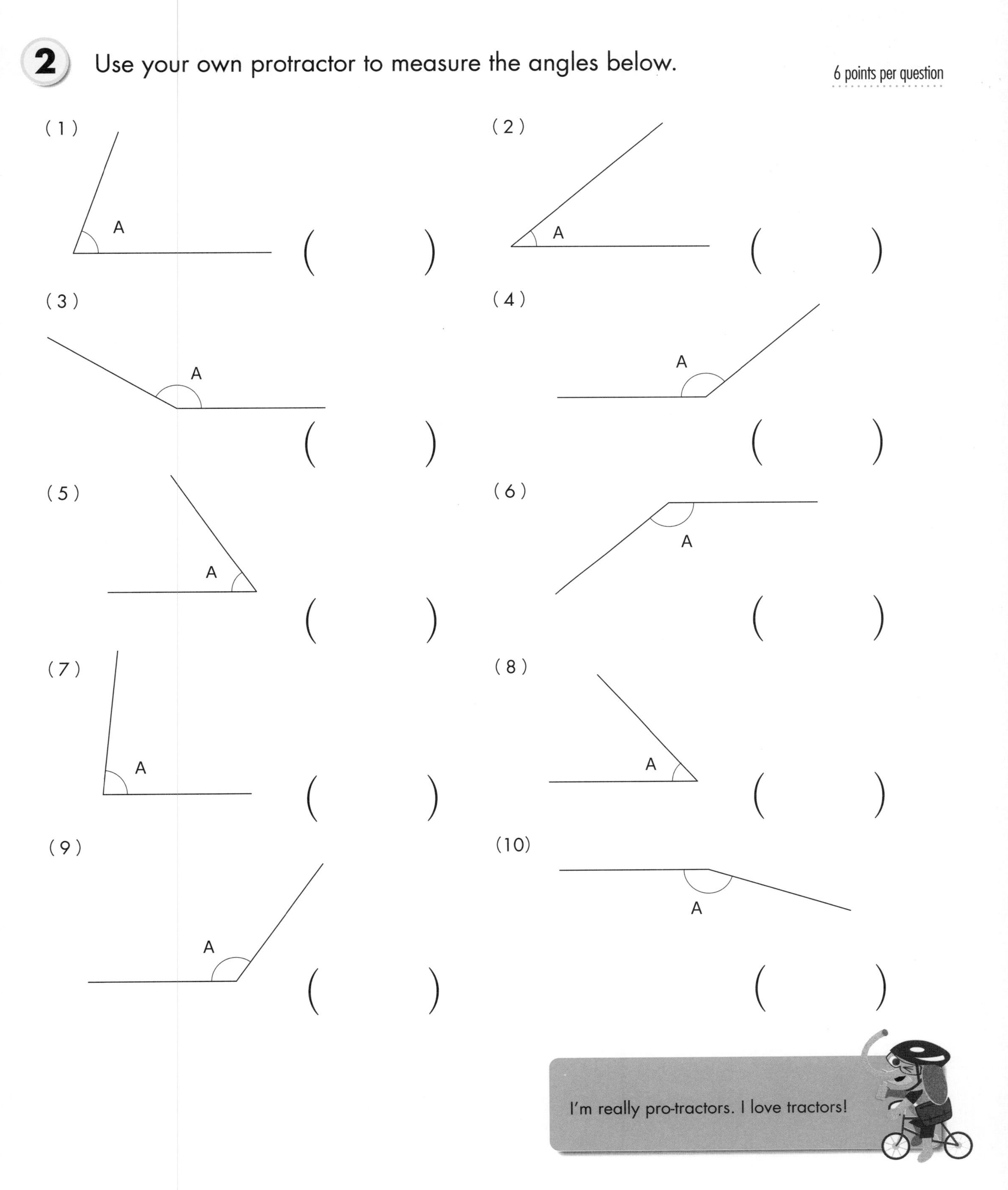

2
Use your own protractor to measure the angles below.
6 points per question
(1)
A
(2)
A
(3)
A
(4)
A
(5)
A
(6)
A
(7)
A
(8)
A
(9)
A
(10)
A
I'm really pro-tractors. I love tractors!

33 Angles & Triangles

Level ★★

Date / / Name

Score /100

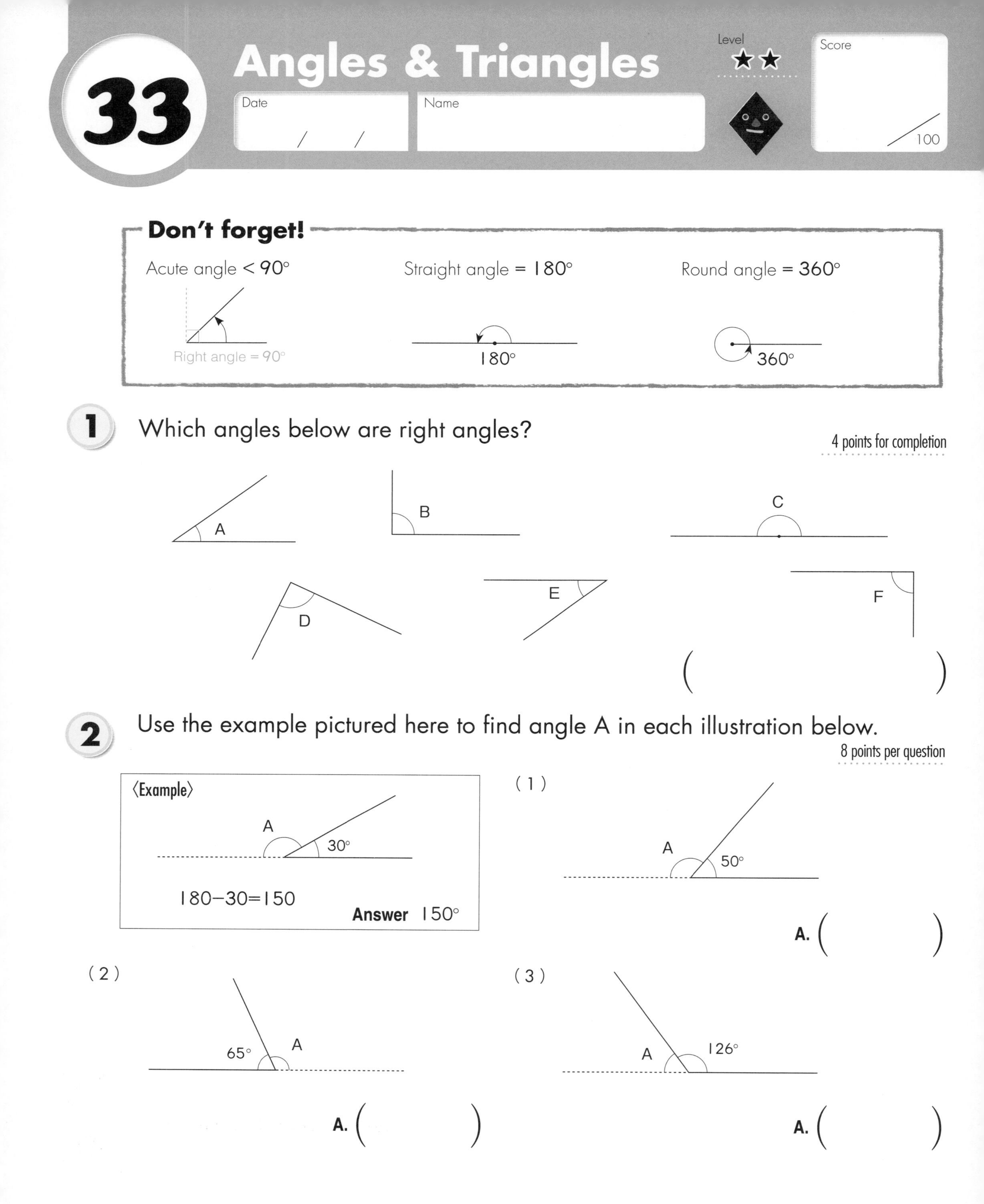

1 Which angles below are right angles?

4 points for completion

( )

2 Use the example pictured here to find angle A in each illustration below.

8 points per question

(1) **A.** ( )

(2) **A.** ( )

(3) **A.** ( )

3 Use the example pictured here to find angle A in each illustration below.

8 points per question

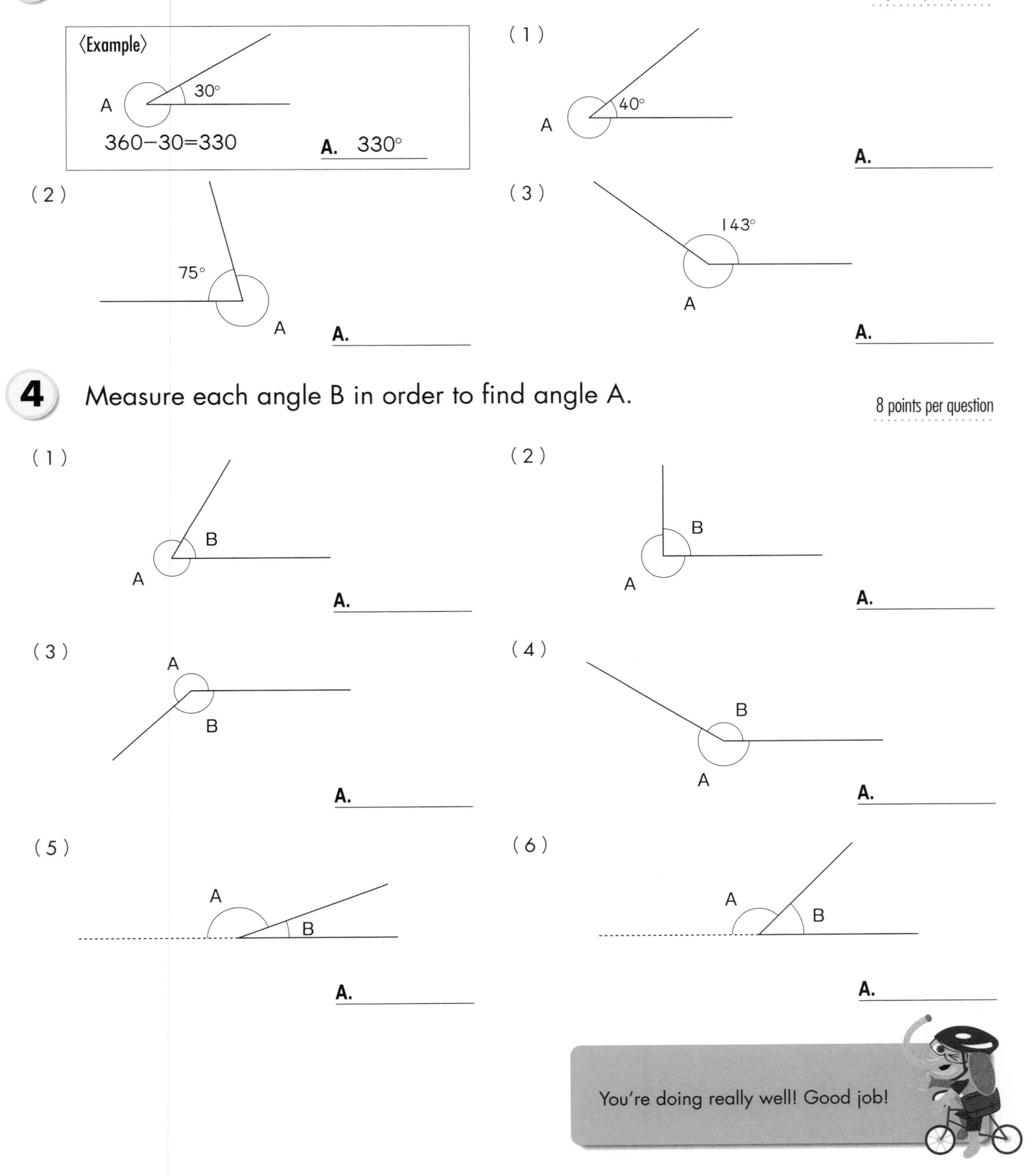

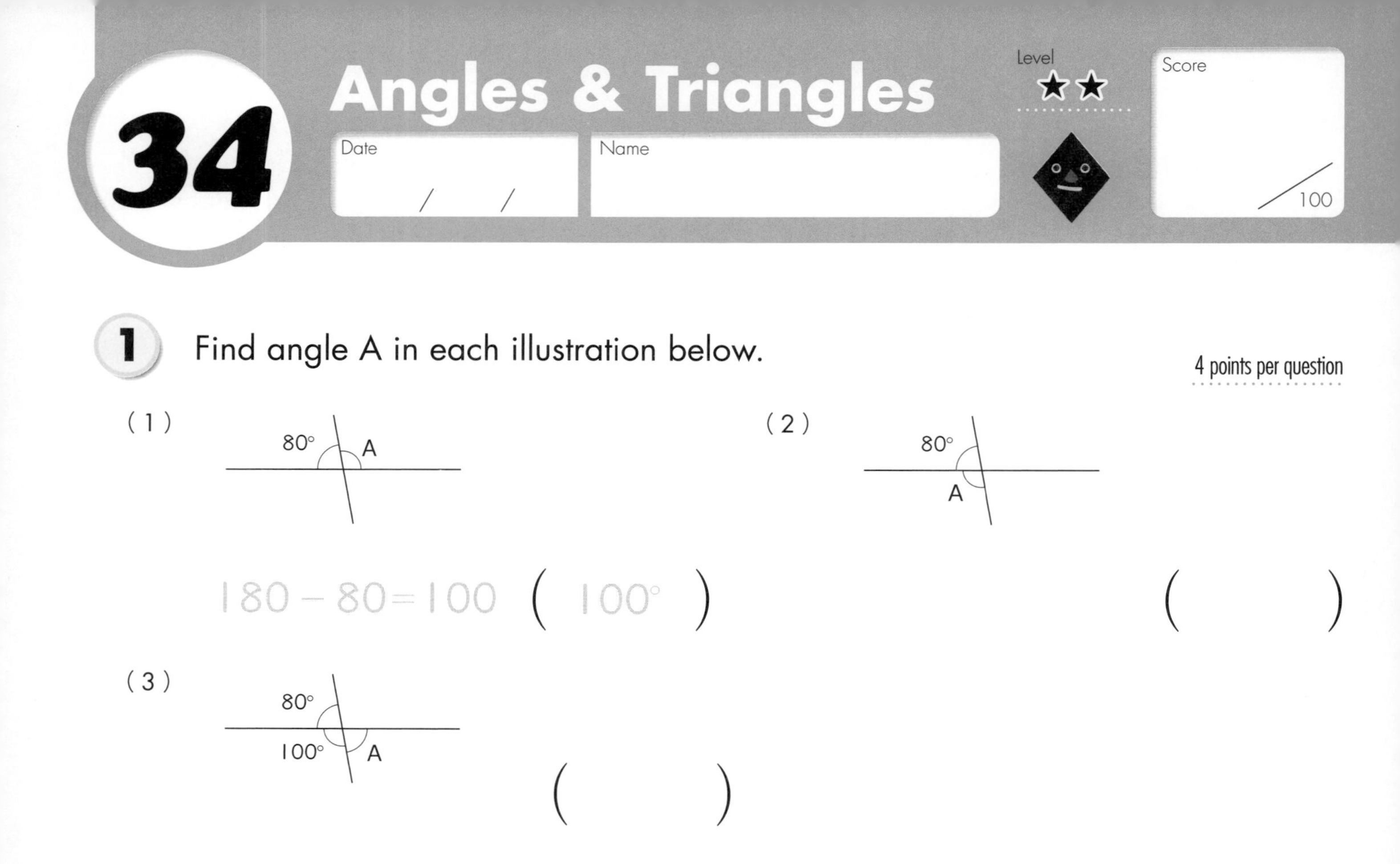

Angles & Triangles

1 Find angle A in each illustration below.

4 points per question

(1) 80° A

$180 - 80 = 100$ (100°)

(2) 80° A

( )

(3) 80° 100° A

( )

2 Use your protractor to measure angles A, B, C and D in the illustration on the right.

4 points per question

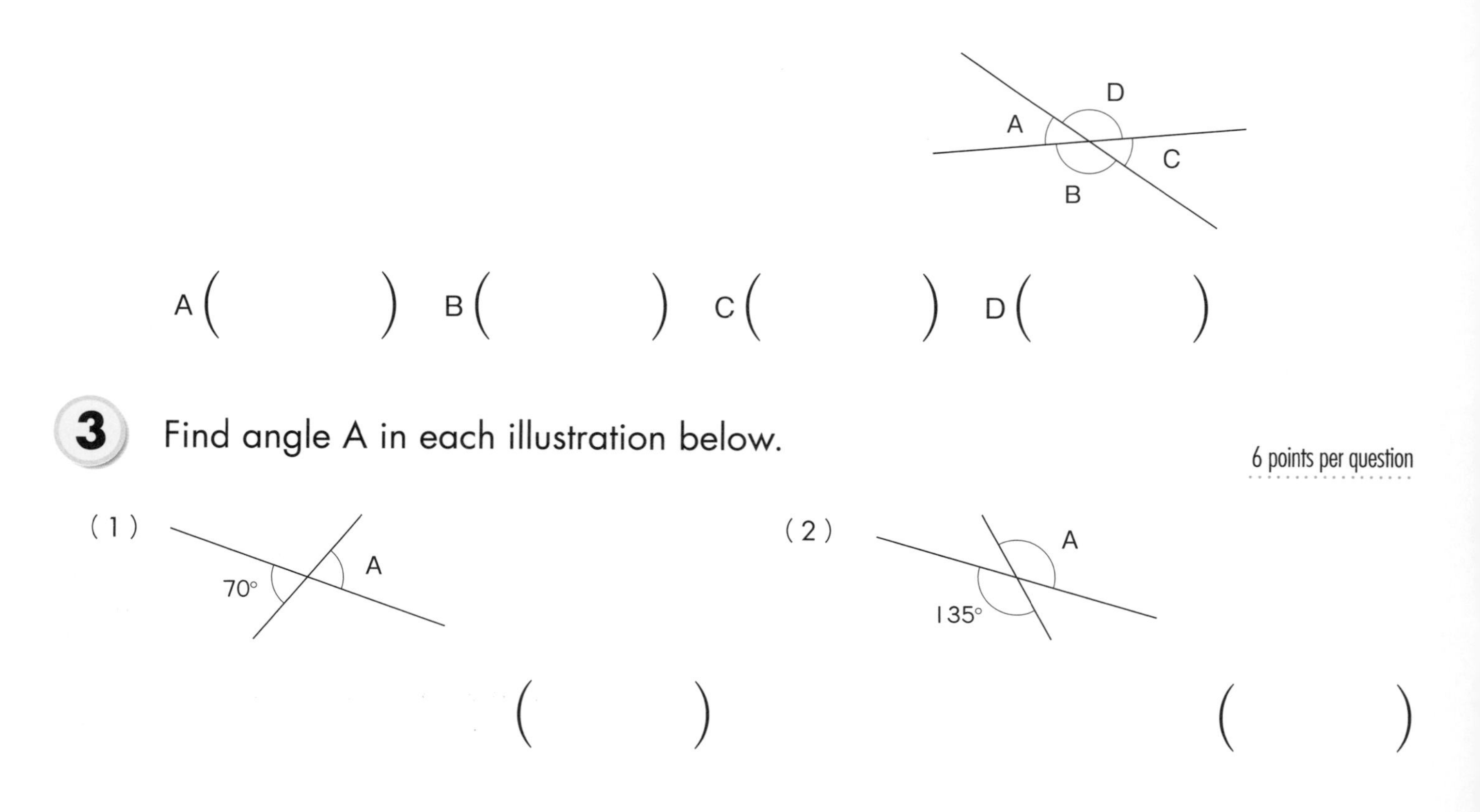

A( ) B( ) C( ) D( )

3 Find angle A in each illustration below.

6 points per question

(1) ( )

(2) ( )

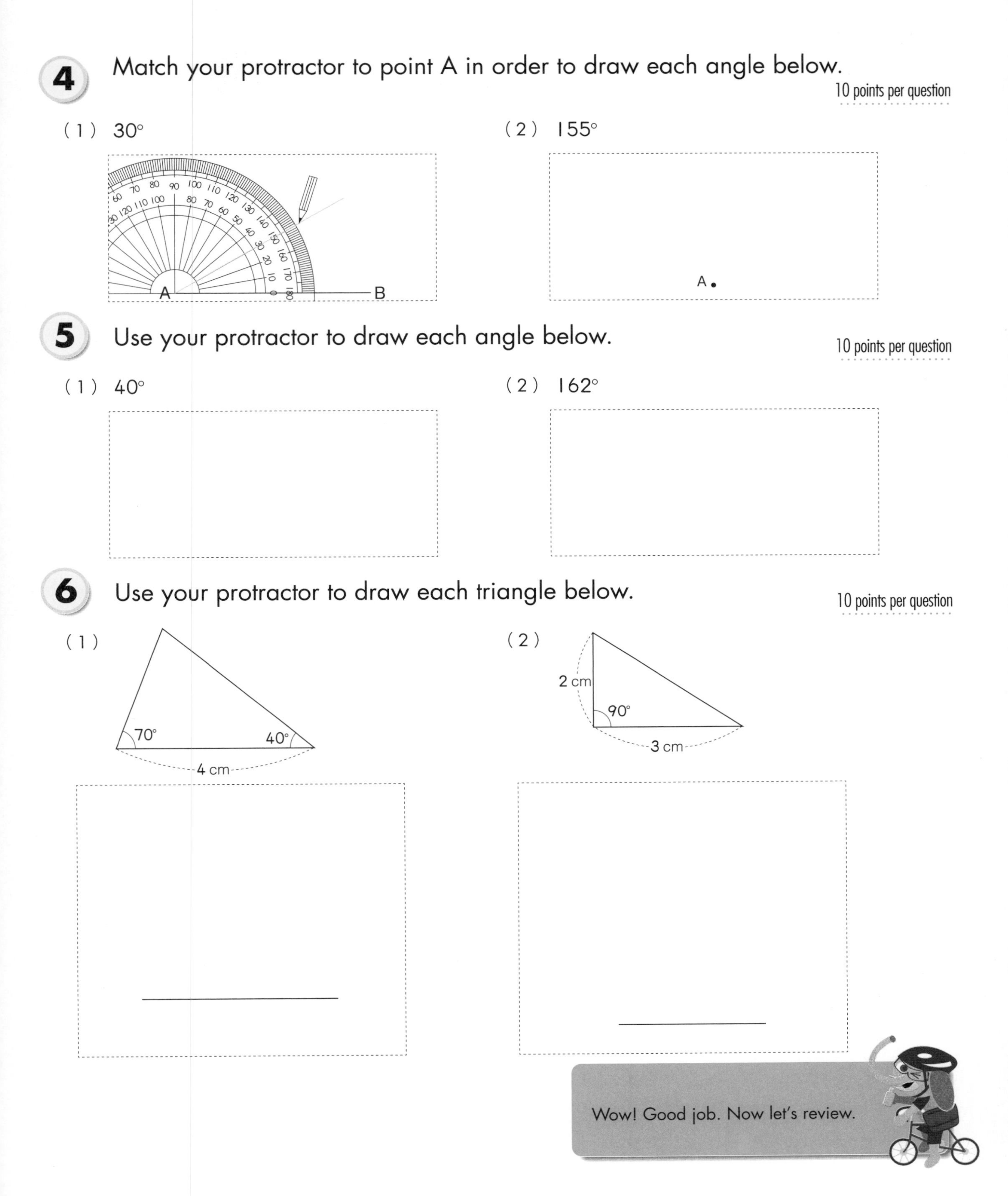

4 Match your protractor to point A in order to draw each angle below.

10 points per question

(1) 30°

(2) 155°

5 Use your protractor to draw each angle below.

10 points per question

(1) 40°

(2) 162°

6 Use your protractor to draw each triangle below.

10 points per question

(1)

(2)

Wow! Good job. Now let's review.

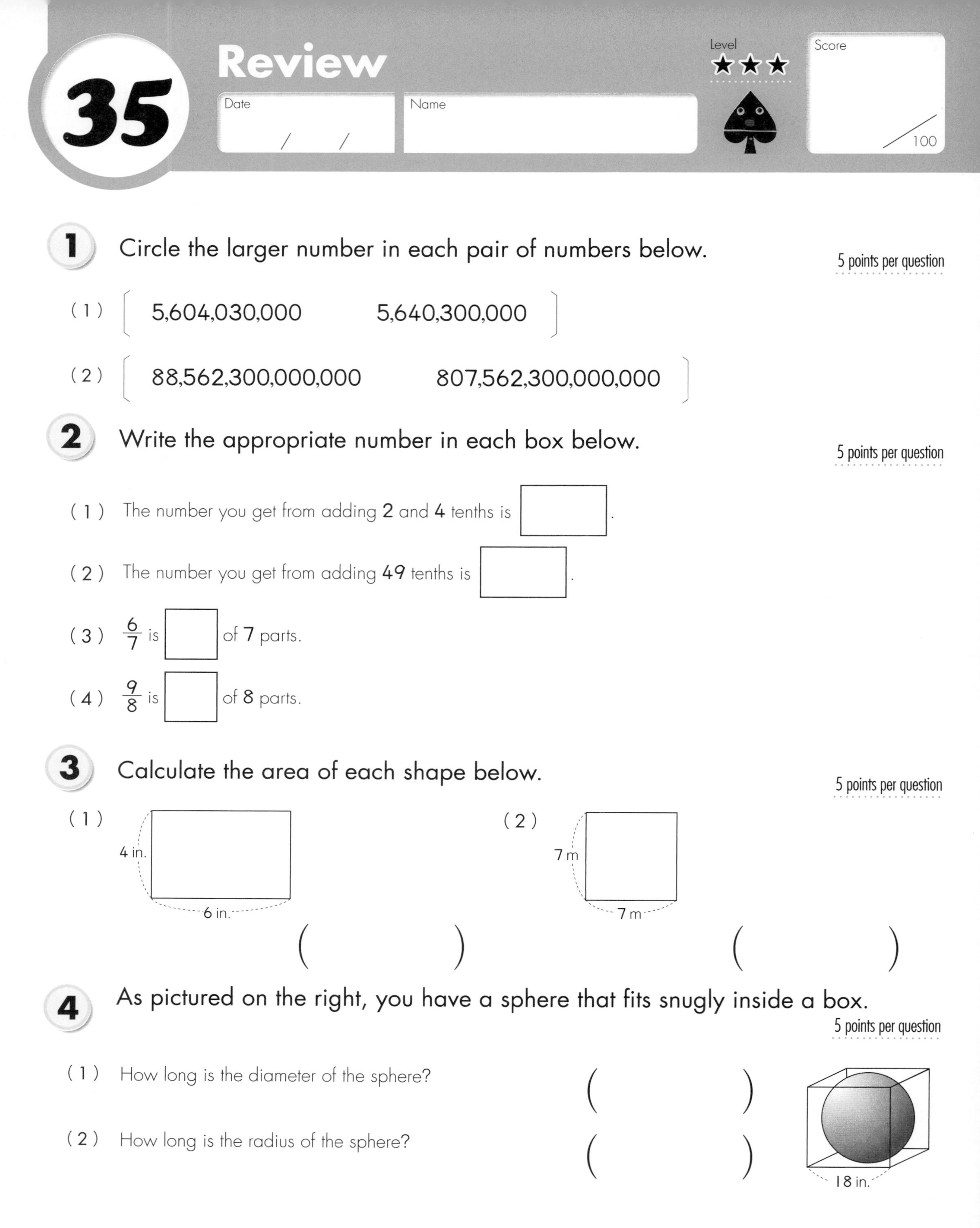

35 Review

Level ★★★

Date / /

Name

Score /100

1 Circle the larger number in each pair of numbers below.

5 points per question

(1) [5,604,030,000 5,640,300,000]

(2) [88,562,300,000,000 807,562,300,000,000]

2 Write the appropriate number in each box below.

5 points per question

(1) The number you get from adding 2 and 4 tenths is ☐.

(2) The number you get from adding 49 tenths is ☐.

(3) $\frac{6}{7}$ is ☐ of 7 parts.

(4) $\frac{9}{8}$ is ☐ of 8 parts.

3 Calculate the area of each shape below.

5 points per question

(1)

( )

(2)

( )

4 As pictured on the right, you have a sphere that fits snugly inside a box.

5 points per question

(1) How long is the diameter of the sphere? ( )

(2) How long is the radius of the sphere? ( )

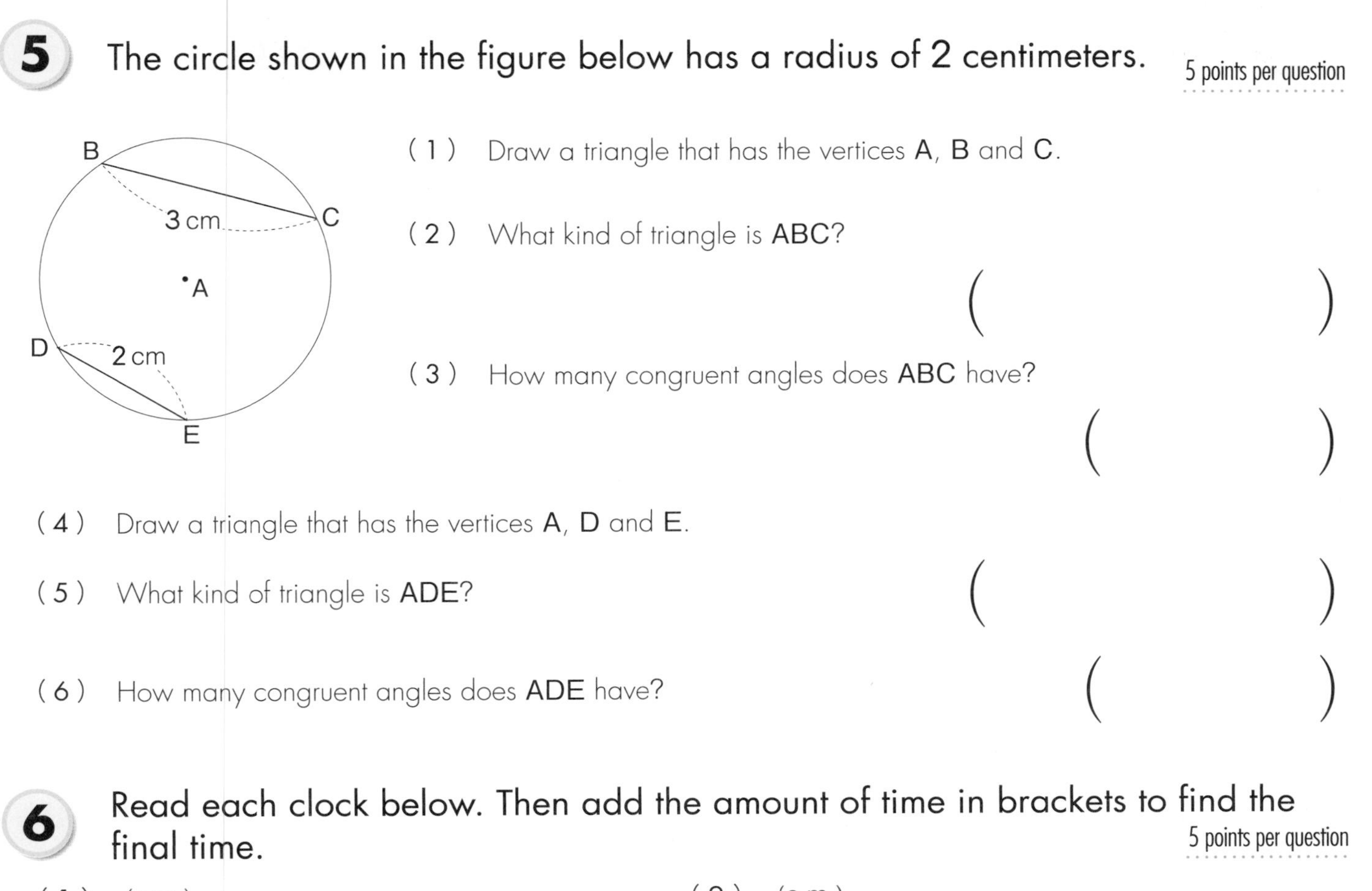

5 The circle shown in the figure below has a radius of 2 centimeters.

5 points per question

(1) Draw a triangle that has the vertices A, B and C.

(2) What kind of triangle is ABC? (   )

(3) How many congruent angles does ABC have? (   )

(4) Draw a triangle that has the vertices A, D and E.

(5) What kind of triangle is ADE? (   )

(6) How many congruent angles does ADE have? (   )

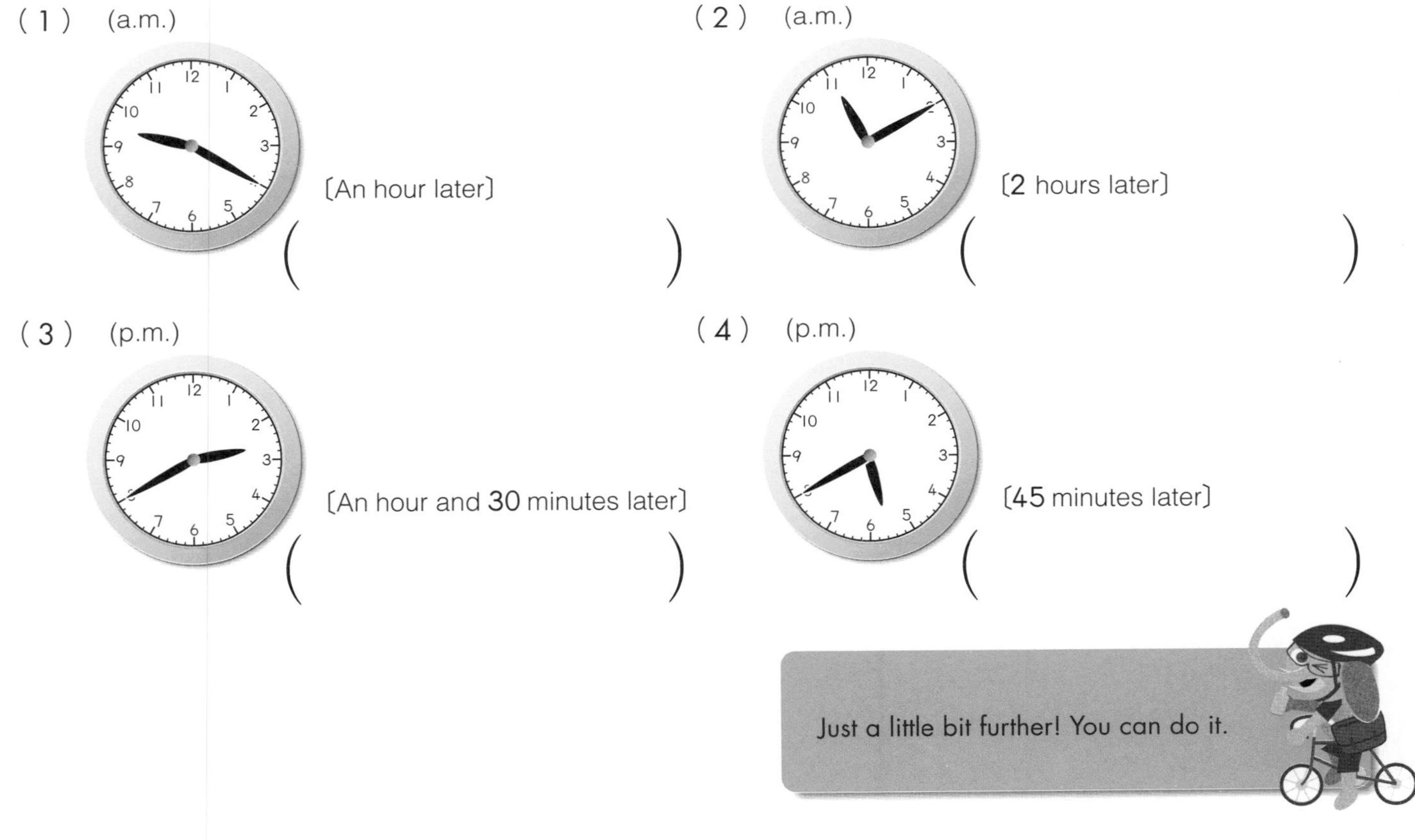

6 Read each clock below. Then add the amount of time in brackets to find the final time.

5 points per question

(1) (a.m.) 〔An hour later〕 (   )

(2) (a.m.) 〔2 hours later〕 (   )

(3) (p.m.) 〔An hour and 30 minutes later〕 (   )

(4) (p.m.) 〔45 minutes later〕 (   )

Just a little bit further! You can do it.

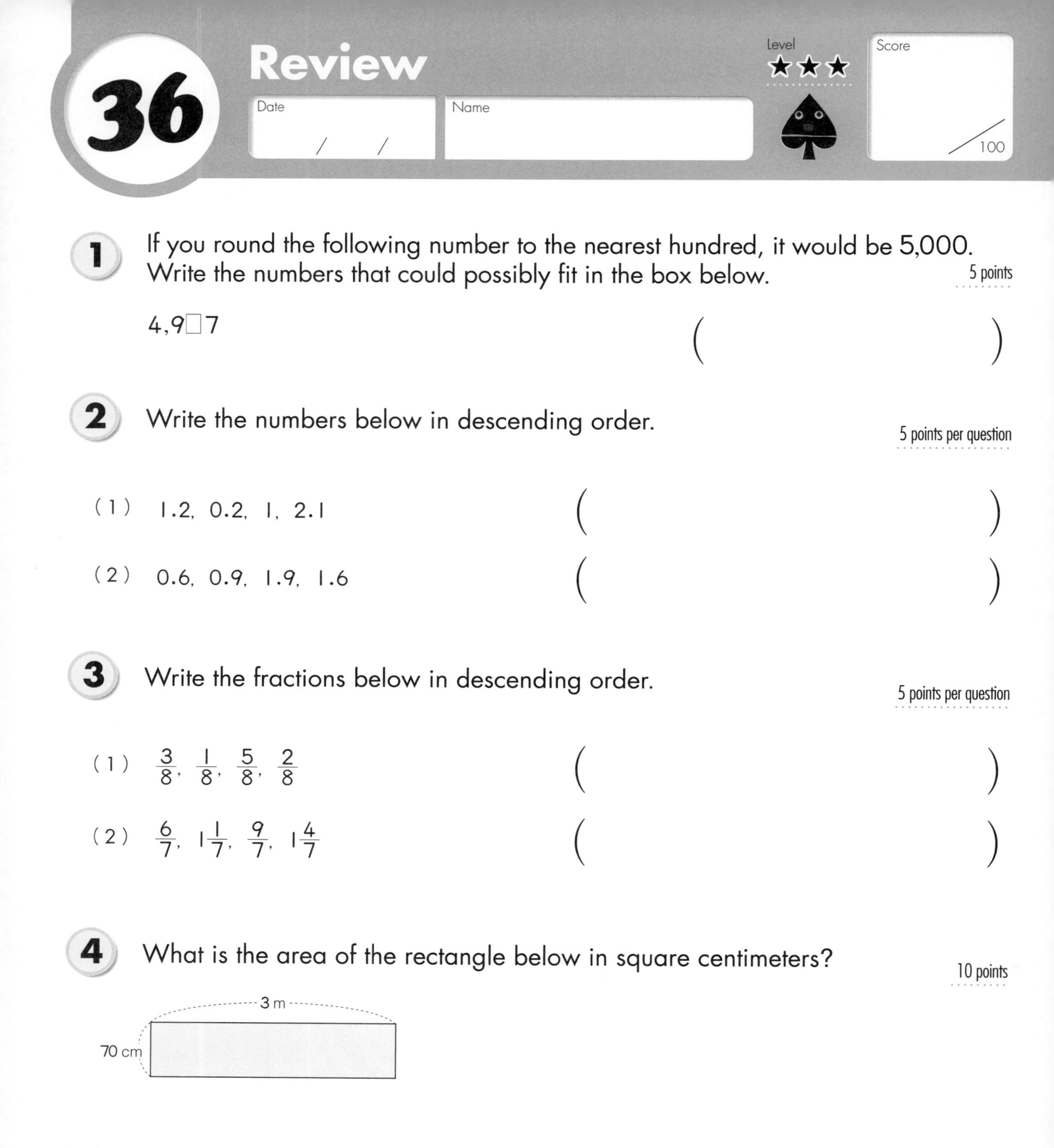

36 Review

Level ★★★

Score /100

Date / /

Name

1 If you round the following number to the nearest hundred, it would be 5,000. Write the numbers that could possibly fit in the box below. 5 points

4,9□7 ( )

2 Write the numbers below in descending order. 5 points per question

(1) 1.2, 0.2, 1, 2.1 ( )

(2) 0.6, 0.9, 1.9, 1.6 ( )

3 Write the fractions below in descending order. 5 points per question

(1) $\frac{3}{8}, \frac{1}{8}, \frac{5}{8}, \frac{2}{8}$ ( )

(2) $\frac{6}{7}, 1\frac{1}{7}, \frac{9}{7}, 1\frac{4}{7}$ ( )

4 What is the area of the rectangle below in square centimeters? 10 points

( )

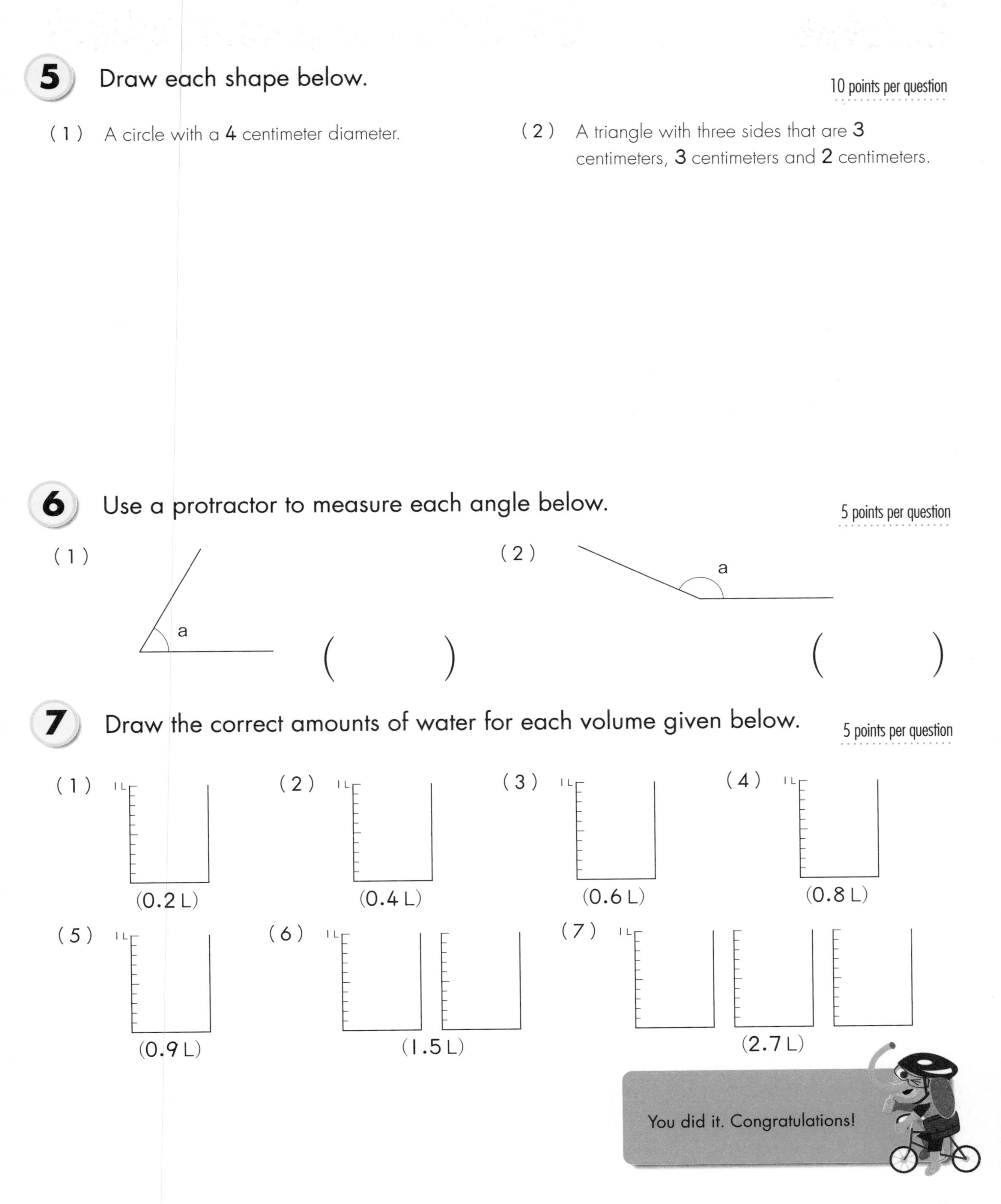

5 Draw each shape below.

10 points per question

(1) A circle with a 4 centimeter diameter.

(2) A triangle with three sides that are 3 centimeters, 3 centimeters and 2 centimeters.

6 Use a protractor to measure each angle below.

5 points per question

(1) a ( )

(2) a ( )

7 Draw the correct amounts of water for each volume given below.

5 points per question

(1) 1 L (0.2 L)

(2) 1 L (0.4 L)

(3) 1 L (0.6 L)

(4) 1 L (0.8 L)

(5) 1 L (0.9 L)

(6) 1 L (1.5 L)

(7) 1 L (2.7 L)

You did it. Congratulations!

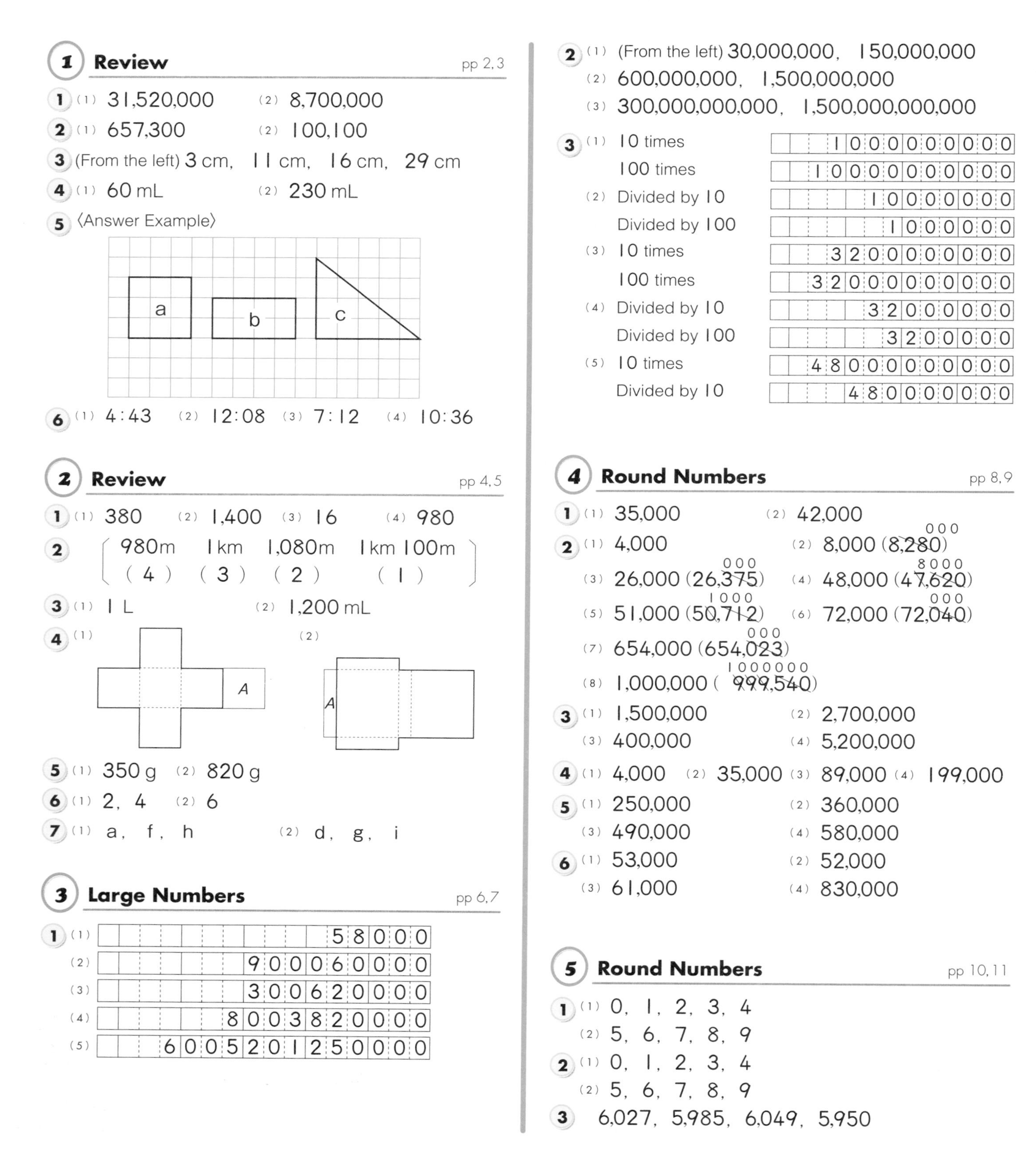

1 Review pp 2, 3

1 (1) 31,520,000 (2) 8,700,000

2 (1) 657,300 (2) 100,100

3 (From the left) 3 cm, 11 cm, 16 cm, 29 cm

4 (1) 60 mL (2) 230 mL

5 ⟨Answer Example⟩

6 (1) 4:43 (2) 12:08 (3) 7:12 (4) 10:36

2 Review pp 4, 5

1 (1) 380 (2) 1,400 (3) 16 (4) 980

2 980m (4) 1km (3) 1,080m (2) 1km 100m (1)

3 (1) 1 L (2) 1,200 mL

4 (1) (2)

5 (1) 350 g (2) 820 g

6 (1) 2, 4 (2) 6

7 (1) a, f, h (2) d, g, i

3 Large Numbers pp 6, 7

1 (1) 58000
(2) 9000060000
(3) 3006200000
(4) 80038200000
(5) 6005201250000

2 (1) (From the left) 30,000,000, 150,000,000
(2) 600,000,000, 1,500,000,000
(3) 300,000,000,000, 1,500,000,000,000

3 (1) 10 times 1000000000
100 times 10000000000
(2) Divided by 10 10000000
Divided by 100 1000000
(3) 10 times 3200000000
100 times 32000000000
(4) Divided by 10 32000000
Divided by 100 3200000
(5) 10 times 48000000000
Divided by 10 480000000

4 Round Numbers pp 8, 9

1 (1) 35,000 (2) 42,000

2 (1) 4,000 (2) 8,000 (8,280)
(3) 26,000 (26,375) (4) 48,000 (47,620)
(5) 51,000 (50,712) (6) 72,000 (72,040)
(7) 654,000 (654,023)
(8) 1,000,000 (999,540)

3 (1) 1,500,000 (2) 2,700,000
(3) 400,000 (4) 5,200,000

4 (1) 4,000 (2) 35,000 (3) 89,000 (4) 199,000

5 (1) 250,000 (2) 360,000
(3) 490,000 (4) 580,000

6 (1) 53,000 (2) 52,000
(3) 61,000 (4) 830,000

5 Round Numbers pp 10, 11

1 (1) 0, 1, 2, 3, 4
(2) 5, 6, 7, 8, 9

2 (1) 0, 1, 2, 3, 4
(2) 5, 6, 7, 8, 9

3 6,027, 5,985, 6,049, 5,950

4 (1) 85, 86, 87, 88, 89, 90, 91, 92, 93, 94
(2) 145, 146, 147, 148, 149, 150, 151, 152, 153, 154
(3) 1,255, 1,256, 1,257, 1,258, 1,259, 1,260, 1,261, 1,262, 1,263, 1,264

5 (1) From 1,650 to 1,749
(2) From 2,950 to 3,049
(3) From 48,500 to 49,499
(4) From 59,500 to 60,499
(5) From 525,000 to 534,999

6 (1) From 885 to 894
(2) From 4,250 to 4,349
(3) From 9,050 to 9,149
(4) From 14,500 to 15,499
(5) From 71,500 to 72,499

7 (1) 55,499 (2) 54,500

6 Round Numbers

pp 12, 13

1 (1) 1,240 + 3,568
1,200 + 3,600 = 4,800
(2) 2,087 + 326
2,100 + 300 = 2,400
(3) 3,472 + 8,735
3,500 + 8,700 = 12,200
(4) 452 + 8,117
500 + 8,100 = 8,600
(5) 8,608 − 6,594
8,600 − 6,600 = 2,000
(6) 7,721 − 945
7,700 − 1,000 = 6,700
(7) 6,890 − 4,683
6,900 − 4,700 = 2,200
(8) 5,667 − 875
5,700 − 900 = 4,800

2 (1) 34,326 + 25,587
34,000 + 26,000 = 60,000
(2) 18,752 + 3,624
19,000 + 4,000 = 23,000
(3) 75,819 + 36,490
76,000 + 36,000 = 112,000
(4) 5,921 + 86,418
6,000 + 86,000 = 92,000
(5) 98,107 − 76,805
98,000 − 77,000 = 21,000
(6) 88,055 − 9,104
89,000 − 9,000 = 80,000
(7) 7,608 + 3,452 + 4,548
8,000 + 3,000 + 5,000 = 16,000
(8) 8,953 − 2,421 − 3,784
9,000 − 2,000 − 4,000 = 3,000

3 (1) 62,543 + 36,875
60,000 + 40,000 = 100,000
(2) 49,803 + 222,675
50,000 + 220,000 = 270,000
(3) 96,873 − 50,032
100,000 − 50,000 = 50,000
(4) 572,156 − 431,890
570,000 − 430,000 = 140,000
(5) 680,289 + 4,569,741 − 58,040
680,000 + 4,570,000 − 60,000 = 5,190,000

4 (1) 348 + 82
350 + 80 = 430
(2) 352 + 3,135
400 + 3,100 = 3,500
(3) 4,513 + 6,462
4,500 + 6,500 = 11,000
(4) 34,910 + 8,294
35,000 + 8,000 = 43,000
(5) 73,404 − 52,783
73,000 − 53,000 = 20,000
(6) 149,321 + 323,584
150,000 + 320,000 = 470,000
(7) 386,304 − 257,392 + 523,608
390,000 − 260,000 + 520,000 = 650,000

7 Fractions

pp 14, 15

1 (1) $\frac{2}{5}$ (2) $\frac{3}{5}$ (3) 5 (4) 5 (5) 4 (6) 5
(7) 1 (8) 5

2 (1) (From the left) $\frac{2}{6}$, $\frac{5}{6}$ (2) $\frac{2}{7}$, $\frac{4}{7}$
(3) $\frac{3}{10}$, $\frac{7}{10}$

3 (1) $\frac{4}{6}$ (2) $\frac{6}{6}$, 1 (3) $\frac{6}{7}$ (4) 7
(5) 6 (6) $\frac{10}{10}$, 1

8 Fractions
pp 16, 17

1 (1) 4 (2) 5 (3) $\frac{5}{10}$ (4) 10 (5) $\frac{9}{10}$ (6) 1

2 (1) 2 (2) 3 (3) 9 (4) 9 (5) $\frac{9}{9}$ (6) 1

3 (1) $\frac{2}{9}$ (2) $\frac{8}{9}$ (3) $\frac{9}{9}$
(4) $\frac{5}{8}$ (5) $\frac{6}{8}$ (6) 1
(7) 1 (8) $\frac{6}{7}$ (9) $\frac{7}{7}$

4 (1) $\frac{3}{4}$ (2) $\frac{4}{5}$ (3) $\frac{7}{7}$ (1) (4) $\frac{6}{8}$

9 Fractions
pp 18, 19

1 (1) $\frac{4}{3}$ L $= 1\frac{1}{3}$ L (2) $\frac{7}{4}$ L $= 1\frac{3}{4}$ L
(3) $\frac{9}{4}$ L $= 2\frac{1}{4}$ L (4) $\frac{13}{5}$ L $= 2\frac{3}{5}$ L

2 (1) $\frac{5}{4}$ m $= 1\frac{1}{4}$ m (2) $\frac{7}{6}$ m $= 1\frac{1}{6}$ m
(3) $\frac{11}{5}$ m $= 2\frac{1}{5}$ m

3 (Proper fractions) $\frac{6}{11}$, $\frac{13}{18}$, $\frac{20}{23}$
(Improper fractions) $\frac{7}{6}$, $\frac{15}{15}$, $\frac{23}{18}$
(Mixed fractions) $1\frac{2}{3}$, $3\frac{15}{17}$, $4\frac{13}{25}$

4 (1) (From the left) $\frac{2}{6}$, $\frac{5}{6}$, $\frac{7}{6}$, $\frac{11}{6}$
(2) $\frac{6}{7}$, $\frac{11}{7}$, $\frac{15}{7}$, $\frac{19}{7}$

10 Fractions
pp 20, 21

1 (From the left) $\frac{3}{6}$, $1\frac{2}{6}$, $1\frac{4}{6}$, $2\frac{1}{6}$, $2\frac{5}{6}$

2 (1) $\frac{6}{5}$, $1\frac{1}{5}$ (2) $\frac{11}{7}$, $1\frac{4}{7}$
(3) 7 (4) 5

3 (1) $\frac{6}{5}$ (2) $\frac{9}{7}$ (3) $1\frac{1}{4}$
(4) $1\frac{2}{9}$

4 (1) $1\frac{1}{3}$ (2) $1\frac{2}{5}$ (3) 1 (4) $2\frac{1}{3}$
(5) 2 (6) $1\frac{5}{6}$ (7) $1\frac{3}{8}$ (8) $1\frac{5}{7}$
(9) $2\frac{1}{4}$

5 (1) $\frac{5}{4}$ (2) $\frac{4}{3}$ (3) $\frac{7}{5}$ (4) $\frac{9}{7}$
(5) $\frac{9}{4}$ (6) $\frac{8}{3}$

6 (1) $\frac{9}{5}$, $1\frac{2}{5}$, $\frac{6}{5}$, $\frac{3}{5}$
(2) $1\frac{4}{9}$, $\frac{11}{9}$, $1\frac{1}{9}$, $\frac{7}{9}$

11 Fractions
pp 22, 23

1 b : $\frac{2}{3}$ c : (From the left) $\frac{2}{4}$, $\frac{3}{4}$ d : $\frac{2}{5}$
e : $\frac{2}{6}$, $\frac{4}{6}$ f : $\frac{1}{7}$, $\frac{4}{7}$, $\frac{6}{7}$ g : $\frac{2}{8}$, $\frac{6}{8}$
h : $\frac{3}{9}$, $\frac{6}{9}$ i : $\frac{4}{10}$, $\frac{9}{10}$

2 (1) $\frac{2}{4}$, $\frac{3}{6}$, $\frac{4}{8}$, $\frac{5}{10}$
(2) $\frac{2}{6}$, $\frac{3}{9}$

3 (1) $\frac{3}{4}$ (2) $\frac{7}{9}$ (3) $\frac{1}{3}$ (4) $\frac{1}{5}$ (5) $\frac{2}{3}$ (6) $\frac{3}{4}$
(7) $\frac{7}{8}$ (8) $\frac{4}{5}$

4 (1) $\frac{8}{9}$, $\frac{7}{9}$, $\frac{4}{9}$, $\frac{2}{9}$
(2) $\frac{5}{6}$, $\frac{5}{7}$, $\frac{5}{8}$, $\frac{5}{9}$
(3) $\frac{6}{5}$, $\frac{6}{7}$, $\frac{6}{8}$, $\frac{6}{11}$

12 Decimals
pp 24, 25

1 (1) 0.1 (2) 0.2 (3) 0.6 (4) 0.9
(5) 1.1 (6) 1.2 (7) 1.5 (8) 1.7
(9) 2.1 (10) 2.3

2 (1) 0.1 (L), 100 (mL) (2) 0.8 (L), 800 (mL)

3 (1) 1.4 (L), 1,400 (mL)
(2) 2.5 (L), 2,500 (mL)

4 (1) 1L (2) 1L (3) 1L (4) 1L (5) 1L (6) 1L

13 Decimals
pp 26, 27

1 (1) 0.1 (2) 0.2 (3) 0.7 (4) 1.1
(5) 1.5 (6) 1.8

2 (1) 4.5 (2) 8.4

3 A 0.8 B 1.6 C 4.3 D 9.1 E 12.5

4 (1) 0.6 (cm), 6 (mm)
(2) 0.9 (cm), 9 (mm)

5 (1) 1.4 (cm), 1 (cm) 4 (mm)
(2) 5.8 (cm), 5 (cm) 8 (mm)

6

Advice
Please ask your parents to check if your answer is correct.

14 Decimals
pp 28, 29

1 (1) 26, 0, 5, 9
(2) 2.6, 0.5, 3.9, 1.9

2 (1) 3, 6 (2) 4, 5 (3) 7 (4) 6.2
(5) 9.7

3 (1) 1.5 (2) 10 (3) 15

4 (1) 9 (2) 19 (3) 1 (4) 21
(5) 2.7 (6) 2.7 (7) 3.4 (8) 5.8

15 Decimals
pp 30, 31

1 (1) (From the left) 0.1, 0.6, 1.4, 1.9
(2) 0.2, 1.1, 2.2, 2.8
(3) 0.4, 1.3, 2.5, 3.9
(4) 0.3, 1.7, 2.6, 4.4

2 A B C D
0 1 2 3

3 (1) 0.7 (2) 1.1 (3) 2.6
(4) 2.2 (5) 3.1 (6) 3.1

4 (1) 1.1 (2) 4.3 (3) 7 (4) 10

5 (1) 1.8, 1.5, 0.8, 0.5
(2) 9.1, 1.9, 1.6, 0.9

16 Volume
pp 32, 33

1 (1) 1 in.^{3} (2) 2 in.^{3} (3) 2 in.^{3} (4) 3 in.^{3}
(5) 3 in.^{3} (6) 4 in.^{3} (7) 4 in.^{3} (8) 6 in.^{3}
(9) 6 in.^{3} (10) 6 in.^{3} (11) 8 in.^{3}

2 (1) 4 in.^{3} (2) 10 in.^{3} (3) 6 in.^{3} (4) 9 in.^{3}
(5) 7 in.^{3} (6) 15 in.^{3} (7) 16 in.^{3} (8) 32 in.^{3}

17 Volume
pp 34, 35

1 (1) 1 cm^3 (2) 2 cm^3 (3) 2 cm^3 (4) 3 cm^3
(5) 3 cm^3 (6) 4 cm^3 (7) 4 cm^3 (8) 6 cm^3
(9) 6 cm^3 (10) 6 cm^3 (11) 8 cm^3

2 (1) 5 cm^3 (2) 8 cm^3 (3) 9 cm^3 (4) 6 cm^3
(5) 6 cm^3 (6) 12 cm^3 (7) 24 cm^3 (8) 40 cm^3

18 Capacity
pp 36, 37

1 (1) 1 (2) 2 (3) 3 (4) 4

2 (1) 2 (2) 4 (3) 6
(4) 1 (5) $\frac{1}{2}$ (6) 2

3 (1) 8 (2) 16 (3) 40
(4) 1 (5) 3 (6) 4

4 (1) 16 oz, 2 cups
(2) 24 oz, 3 cups
(3) 12 oz, $1\frac{1}{2}$ cups

19 Area
pp 38, 39

1 (1) $1 \times 1 = 1$ 1 in.^{2}
(2) $3 \times 1 = 3$ 3 in.^{2}
(3) $2 \times 2 = 4$ 4 in.^{2}
(4) $3 \times 2 = 6$ 6 in.^{2}
(5) $3 \times 3 = 9$ 9 in.^{2}
(6) $9 \times 3 = 27$ 27 in.^{2}

2 (1) $3 \times 2 = 6$ — 6 ft.^2
(2) $5 \times 4 = 20$ — 20 ft.^2
(3) $4 \times 4 = 16$ — 16 ft.^2
(4) $3 \times 8 = 24$ — 24 ft.^2
(5) $7 \times 7 = 49$ — 49 ft.^2
(6) $10 \times 6 = 60$ — 60 ft.^2
(7) $12 \times 5 = 60$ — 60 ft.^2
(8) $8 \times 15 = 120$ — 120 ft.^2

20 Area

pp 40, 41

1 (1) $1 \times 1 = 1$ — 1 cm^2
(2) $2 \times 1 = 2$ — 2 cm^2
(3) $2 \times 2 = 4$ — 4 cm^2
(4) $2 \times 3 = 6$ — 6 cm^2
(5) $3 \times 3 = 9$ — 9 cm^2
(6) $4 \times 3 = 12$ — 12 cm^2
(7) $6 \times 6 = 36$ — 36 cm^2
(8) $7 \times 2 = 14$ — 14 cm^2
(9) $5 \times 9 = 45$ — 45 cm^2

2 (1) $3 \times 2 = 6$ — 6 m^2
(2) $4 \times 3 = 12$ — 12 m^2
(3) $5 \times 5 = 25$ — 25 m^2
(4) $3 \times 7 = 21$ — 21 m^2
(5) $7 \times 7 = 49$ — 49 m^2
(6) $10 \times 5 = 50$ — 50 m^2
(7) $12 \times 6 = 72$ — 72 m^2
(8) $7 \times 14 = 98$ — 98 m^2

21 Area

pp 42, 43

1 (1) 1 m = 100 cm, $100 \times 40 = 4{,}000$
Ans. $4{,}000 \text{ cm}^2$
(2) 1 m 50 cm = 150 cm, $60 \times 150 = 9{,}000$
Ans. $9{,}000 \text{ cm}^2$

2 (1) $100 \times 100 = 10{,}000$ **Ans.** $10{,}000 \text{ cm}^2$
(2) 100 cm = 1 m, $1 \times 1 = 1$
Ans. 1 m^2
(3) $200 \times 200 = 40{,}000$ **Ans.** $40{,}000 \text{ cm}^2$
(4) 200 cm = 2 m, $2 \times 2 = 4$
Ans. 4 m^2

3 (1) 2 m = 200 cm, $200 \times 50 = 10{,}000$
$10{,}000 \text{ cm}^2 = 1 \text{ m}^2$ **Ans.** 1 m^2
〔Also, 50 cm = 0.5 m, $2 \times 0.5 = 1$〕
(2) 5 m = 500 cm, 2 m 40 cm = 240 cm
$240 \times 500 = 120{,}000$
$120{,}000 \text{ cm}^2 = 12 \text{ m}^2$ **Ans.** 12 m^2
〔Also, 2 m 40 cm = 2.4 m, $2.4 \times 5 = 12$〕

4 1 km = 1,000 m, $1{,}000 \times 300 = 300{,}000$
Ans. $300{,}000 \text{ m}^2$

5 (1) 1 km = 1,000 m, $800 \times 1{,}000 = 800{,}000$
Ans. $800{,}000 \text{ m}^2$
(2) 2 km = 2,000 m, $2{,}000 \times 70 = 140{,}000$
Ans. $140{,}000 \text{ m}^2$

22 Elapsed Time

pp 44, 45

1 (1) 1 (2) 2 (3) 3 (4) 60
(5) 1 (6) 60 (7) 120 (8) 2
(9) 120 (10) 90

2 (1) a.m. (2) p.m. (3) 12 (4) 12
(5) 12, 12 (6) 24

23 Elapsed Time

pp 46, 47

1 (1) midnight (12:00 a.m.)
(2) midnight (12:00 a.m.) (3) 12 (4) 2
(5) 7 (6) 7 (7) 10 (8) 11

2 (1) 6:20 a.m. (2) 12:40 p.m.
(3) 11:35 a.m. (4) 1:20 p.m.
(5) 7:40 a.m. (6) 10:24 p.m.

3 (1) 11:00 a.m. (2) noon (12:00 p.m.)
(3) 10:30 a.m. (4) 12:30 a.m.

24 Elapsed Time

pp 48, 49

1 (1) 20 minutes (2) 30 minutes
(3) 25 minutes (4) 20 minutes

2 (1) 1 hour (2) 2 hours
(3) 1 hour 30 minutes
(4) 1 hour 40 minutes

3 (1) 4:00 p.m. (2) 4:10 p.m.
(3) 4:15 p.m. (4) 4:05 p.m.

4 (1) 11:30 a.m. (2) 12:30 p.m.
(3) 5:00 p.m. (4) 4:40 p.m.

25 Elapsed Time

pp 50,51

1 (1) 1 second (2) 5 seconds
(3) 10 seconds (4) 15 seconds
(5) 20 seconds (6) 25 seconds
(7) 30 seconds (8) 40 seconds

2 (1) 1 minute 30 seconds
(2) 1 minute 10 seconds
(3) 1 minute 40 seconds
(4) 1 minute 42 seconds

3 (1) 60 (2) 61 (3) 70 (4) 90
(5) 76 (6) 98 (7) 120 (8) 140
(9) 150 (10) 132 (11) 180 (12) 200
(13) (From the left) 1, 10 (14) 1, 20 (15) 1, 30
(16) 1, 40 (17) 1, 15 (18) 1, 35
(19) 1, 45 (20) 1, 28 (21) 2
(22) 3 (23) 2, 10 (24) 2, 20

26 Circles & Spheres

pp 52,53

1 (1) 4 (2) 1 (3) 8 (4) 2 (5) 3, 5

2 (1) 〈Answer Example〉 (2) 〈Answer Example〉

Center

Center

3 **Advice**
Please ask your parents to check if your answer is correct.

27 Circles & Spheres

pp 54,55

1 (1) 4 in. (2) 4 in. (3) 2 in.
2 (1) 10 in. (2) 5 in. (3) 2.5 in.
3 (1) 6 in. (2) 6 in.
4 (1) 8 cm (2) 10 cm (3) 7 cm
5 8 cm
6 12 cm
7 15 cm

28 Circles & Spheres

pp 56,57

1

A B C

2 (1) 2 (2) 3, 5 (3) 10 (4) 14
3 (1) 9 cm (2) 4 cm 5 mm 〔4.5 cm〕
4 (1) 16 in. (2) 16 in. (3) 8 in.
5 (1) 8 in. (2) 4 in.
6 (1) 6 in. (2) 12 in.

29 Angles & Triangles

pp 58,59

1 (1) Isosceles triangle (2) Equilateral triangle
(3) Isosceles triangle (4) Equilateral triangle

2 (1) a 3 cm b 3 cm
c 5 cm d 3 cm
e 3 cm 5 mm 〔3.5 cm〕 f 2 cm
g 2 cm h 2 cm
i 2 cm
(2) Isosceles triangle…A
Equilateral triangle…C

3 (1) Equilateral triangle…e, g
(2) Isosceles triangle…a, b, d, j

4 (1) Isosceles triangle (2) Equilateral triangle
(3) Isosceles triangle

30 Angles & Triangles

pp 60,61

1 (1) (2)
Use the example to draw the triangles. Then use your ruler to check your drawing.

2 (1) (2)
Use the example and then check your answers with a ruler.

3 Use your ruler to check your answer.
(1), (2) are isosceles triangles.
(3), (4) are equilateral triangles.

4 (1) (3)

E B A 4 cm 3 cm D C

(2) Equilateral triangle
(4) Isosceles triangle

31 Angles & Triangles
pp 62,63

1 (1) b (2) b (3) e (4) d (5) a (6) e

2 (1) a…2 b…3 c…1
(2) a…3 b…1 c…2

3 (1) b, c (2) d, e, f (3) h, i

32 Angles & Triangles
pp 64,65

1 (1) 1° (2) 5° (3) 30° (4) 50° (5) 65°
(6) 83° (7) 120° (8) 160°

2 (1) 70° (2) 40° (3) 150° (4) 140° (5) 55°
(6) 140° (7) 84° (8) 48° (9) 126° (10) 163°

33 Angles & Triangles
pp 66,67

1 B, D, F

2 (1) $180-50=130$ **Ans.** 130°
(2) $180-65=115$ **Ans.** 115°
(3) $180-126=54$ **Ans.** 54°

3 (1) $360-40=320$ **Ans.** 320°
(2) $360-75=285$ **Ans.** 285°
(3) $360-143=217$ **Ans.** 217°

4 (1) $360-60=300$ **Ans.** 300°
(2) $360-90=270$ **Ans.** 270°
(3) $360-138=222$ **Ans.** 222°
(4) $360-149=211$ **Ans.** 211°
(5) $180-20=160$ **Ans.** 160°
(6) $180-45=135$ **Ans.** 135°

34 Angles & Triangles
pp 68,69

1 (1) $180-80=100$ **Ans.** 100°
(2) $180-80=100$ **Ans.** 100°
(3) $180-100=80$ **Ans.** 80°

2 A 40° B 140° C 40° D 140°

3 (1) 70° (2) 135°

4 (1) A 30° B (2) 155° A

5 (1) 40° (2) 162°

6 (1) ~ (4)
Check your answers with a ruler and a protractor.

35 Review
pp 70,71

1 (1) 5,640,300,000
(2) 807,562,300,000,000

2 (1) 2.4 (2) 4.9 (3) 6 (4) 9

3 (1) $4\times6=24$ **Ans.** 24 in.^{2}
(2) $7\times7=49$ **Ans.** 49 m^2

4 (1) 18 in. (2) 9 in.

5 (1) (4) B, C, A, D, E, 3 cm, 2 cm
(2) Isosceles triangle
(3) 2 angles
(5) Equilateral triangle
(6) 3 angles

6 (1) 10:20 a.m. (2) 1:10 p.m.
(3) 4:10 p.m. (4) 6:25 p.m.

36 Review
pp 72,73

1 5, 6, 7, 8, 9

2 (1) 2.1, 1.2, 1, 0.2
(2) 1.9, 1.6, 0.9, 0.6

3 (1) $\frac{5}{8}, \frac{3}{8}, \frac{2}{8}, \frac{1}{8}$
(2) $1\frac{4}{7}, \frac{9}{7}, 1\frac{1}{7}, \frac{6}{7}$

4 3 m = 300 cm
$300\times70=21000$ **Ans.** 21,000 cm^2

5 (1) (4 cm) (2) (3 cm) (3 cm) (2 cm)

6 (1) 60° (2) 155°

7 (1) 1 L (0.2 L) (2) 1 L (0.4 L) (3) 1 L (0.6 L) (4) 1 L (0.8 L)
(5) 1 L (0.9 L) (6) 1 L (1.5 L) (7) 1 L (2.7 L)

